Study Guide

Understanding Statistics in the Behavioral Sciences

NINTH EDITION

Robert R. Pagano

Prepared by

Robert R. Pagano

and

William C. Follette

Australia • Brazil • Japan • Korea • Mexico • Singapore • Spain • United Kingdom • United States

For product information and technology assistance, contact us at
Cengage Learning Customer & Sales Support, 1-800-354-9706

For permission to use material from this text or product, submit all requests online at
www.cengage.com/permissions
Further permissions questions can be emailed to
permissionrequest@cengage.com

ISBN-13: 978-0-495-59656-1
ISBN-10: 0-495-59656-6

Wadsworth
10 Davis Drive
Belmont, CA 94002-3098
USA

Cengage Learning is a leading provider of customized learning solutions with office locations around the globe, including Singapore, the United Kingdom, Australia, Mexico, Brazil, and Japan. Locate your local office at: **www.cengage.com/international**

Cengage Learning products are represented in Canada by Nelson Education, Ltd.

To learn more about Brooks/Cole, visit **www.cengage.com/wadsworth**

Purchase any of our products at your local college store or at our preferred online store **www.ichapters.com**

Printed in Canada
1 2 3 4 5 6 7 12 11 10 09

CONTENTS

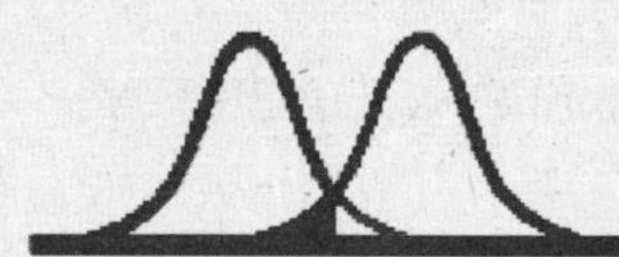

PART ONE OVERVIEW

Chapter 1 STATISTICS AND SCIENTIFIC METHOD 1

PART TWO DESCRIPTIVE STATISTICS

Chapter 2 BASIC MATHEMATICAL AND MEASUREMENT CONCEPTS 13

Chapter 3 FREQUENCY DISTRIBUTIONS 35

Chapter 4 MEASURES OF CENTRAL TENDENCY AND VARIABILITY 63

Chapter 5 THE NORMAL CURVE AND STANDARD SCORES 95

Chapter 6 CORRELATION 117

Chapter 7 LINEAR REGRESSION 147

PART THREE **INFERENTIAL STATISTICS**

Chapter 8 RANDOM SAMPLING AND PROBABILITY 171

Chapter 9 BINOMIAL DISTRIBUTION 195

Chapter 10 INTRODUCTION TO HYPOTHESIS TESTING USING THE SIGN TEST 217

Chapter 11 POWER 243

Chapter 12 SAMPLING DISTRIBUTIONS, SAMPLING DISTRIBUTION OF THE MEAN, THE NORMAL DEVIATE (z) TEST 261

Chapter 13 STUDENT'S t TEST FOR SINGLE SAMPLES 289

Chapter 14 STUDENT'S t TEST FOR CORRELATED AND INDEPENDENT GROUPS 315

Chapter 15 INTRODUCTION TO THE ANALYSIS OF VARIANCE 361

Chapter 16 INTRODUCTION TO TWO-WAY ANALYSIS OF VARIANCE 421

Chapter 17 CHI-SQUARE AND OTHER NONPARAMETRIC TESTS 461

Chapter 18 REVIEW OF INFERENTIAL STATISTICS 509

PREFACE

Statistics uses probability, logic and mathematics as a way of determining whether or not observations made in the real world or laboratory are due to random happenstance or perhaps due to an orderly effect one variable has on another. Separating happenstance, or chance, from cause and effect is the task of science, and statistics is a tool to accomplish that end. Occasionally, data will be so clear that the use of statistical analysis isn't necessary. Occasionally, data will be so garbled that no statistics can meaningfully be applied to it to answer any reasonable question. But we will demonstrate that most often statistics is useful in identifying whether it is legitimate to conclude that an orderly effect has occurred.

It is useful to try to think of statistics as a means of learning a new set of problem solving skills. You will learn new ways to ask questions, new ways to answer them, and a more sophisticated way of interpreting the data you read about in texts, journals, and the newspapers.

Each chapter in this study guide begins by outlining the corresponding chapter in the textbook. Next, there is a section called *Concept Review* that is a programmed learning, "fill in the blanks" exercise. This is a useful way to make sure you understand the basic ideas in the chapter. We recommend you go through this section by covering the answers at the right of the page with a book mark, or ruler, and uncovering each answer after you fill in the blank for that item. Following the concept review, there are sets of *Exercises* which are designed to help you practice the mechanics of solving the problems, recognize what tests to apply to a real world problem and notice any changes in computational procedures required for special cases. A set of *True-False* questions follow the exercise section. Finally, there is a *Self-Quiz* which is presented in a multiple choice format.

In our experience there is a direct, positive relationship between working problems and doing well on this material. We encourage you to work as many problems as you possibly can. When using calculators and computers, there can be a tendency to press the keys and read the answer without having the slightest idea where the answer came from. We hope you won't fall into this trap.

Study Hints

- **Memorize symbols.** A lot of symbols are used in statistics. Don't make the material more difficult than necessary by failing to memorize what the symbols stand for. Treat them as though they were foreign vocabulary. Be able to go quickly from the symbol to the term(s), and from the term(s) to the symbol. There is a section in the accompanying web material that will help you accomplish this goal.
- **Learn the definitions for new terms.** There are a lot of new terms introduced in this course. Part of learning statistics is learning the definitions of these new terms. If you don't know what the new terms mean, it is impossible to do well in this course. Like the symbols, treat the new terms like foreign vocabulary. Be able to go quickly in your head from each new term to its definition and vice versa. There is a section in the accompanying web material that will help you accomplish this goal.
- **Work as many problems as you possibly can.** In my experience there is a direct, positive relationship between working problems and doing well on this material. Be sure

you try to understand the solution. When using calculators and computers, there can be a tendency to press the keys and read the answer without really understanding the solution. We hope you won't fall into this trap. Also, work the problem from beginning to end, rather than just following someone else's solution and telling yourself that you could solve the problem if called upon to do so. Solving a problem from scratch is very different and often more difficult than "understanding" someone else's solution.

- **Don't fall behind.** The material in this course is cumulative. Do not let yourself fall behind. If you do, you will not understand the current material either.
- **Study several times each week, rather than just cramming.** A lot of research has shown that you will learn better and remember more material if you space your learning rather than just cramming for the test.
- **Read the material in the textbook prior to the lecture/discussion covering it.** You can learn a lot just by reading this textbook. Moreover, by reading the appropriate material just prior to when it is covered in class, you can determine the parts that you have difficulty with, and ask appropriate questions when that material is covered by your instructor.
- **Pay attention and think about the material being covered in class.** This advice may seem obvious, but for whatever reason, it is frequently not followed by students. Often times I've had to stop my lecture or discussions to remind students about the importance of paying attention and thinking in class.. I didn't require students to attend my classes, but if they did, I assumed they were interested in learning the material and of course, attention and thinking is prerequisite for learning to take place.
- **Ask the questions you need to ask.** Many of us feel our question is a "dumb" one, and we will be embarrassed because the question will reveal our ignorance to the instructor and the rest of the class. Almost always, the

"dumb" question helps others sitting in the class because they have the same question. Even when this is not true, it is very often the case, that if you don't ask the question, your learning is blocked and stops there, because the answer is necessary for you to continue learning the material. Don't let possible embarrassment hinder your learning. If it doesn't work for you to ask in class, then ask the question via email, or make an appointment with the instructor and ask then.

- **One final point - comparing your answers to ours.** For most of the problems we have used a hand calculator or computer to find the solutions. Depending on how many decimal places you carry your intermediate calculations, you may get slightly different answers than we do. In most cases we have used full calculator or computer accuracy for intermediate calculations (at least five decimal places). In general, you should carry all intermediate calculations to at least two more decimal places than the number of decimal places in the rounded final answer. For example, if you intend to round the final answer to two decimal places, than you should carry all intermediate calculations to at least 4 decimal places. If you follow this policy and your answer does not agree with ours, then you have probably made a calculation error.

Many students have commented that they have found the previous editions of the study guide quite useful. This new edition is very much like the first eight editions, except that it has been changed to cover the material as it is presented in the ninth edition of textbook. We hope you will find the 9th edition helpful in understanding this rather difficult subject matter.

William C. Follette
Robert R. Pagano

Fort Collins, Colorado
September, 2008

Chapter 1

Statistics and Scientific Method

CHAPTER OUTLINE

I. Methods of Acquiring Knowledge

A. Authority. One accepts information as being true because someone who is supposed to know tells you something is true.

B. Rationalism. This method uses reason alone to arrive at knowledge. One analyzes a situation and draws logical conclusions based on the information at hand. The conclusion is not tested empirically to determine if it is correct.

C. Intuition. This is a sudden insight that springs into consciousness all at once as a whole.

D. Scientific method. This method uses reasoning and intuition as a means of formulating an idea of what is true but then relies on objective assessment to verify or deny the validity of the idea.

1. Idea formed and hypothesis made.
2. Experiment designed.
3. Data collected and analyzed using statistics.
4. Hypothesis confirmed, denied or modified.

II. Scientific Research

A. Observational studies. In this research there is no direct experimental manipulation of variables. This technique employs naturalistic observation of events in their real world environment.

1. Correlation. A type of observation where the relationship between two variables is inferred.
2. Parameter estimation. This is when an investigator tries to determine the actual characteristics of the population, based on measuring a subset of the population.

B. True experiments. The investigator attempts to determine if changes in one variable produce changes in another. Only true experiments can legitimately be used to infer causation. In both observational studies and true experiments, statistical analysis is usually employed.

C. Statistical analysis.

1. Descriptive statistics. Analysis is conducted to describe the obtained data.
2. Inferential statistics. Analysis is conducted to make inferences about a population using data obtained from the sample.

CONCEPT REVIEW

There are at least four methods of acquiring knowledge.

When one accepts something as true because of

tradition or because a person of distinction says it is true, one is using the method of (1) ________.	(1) authority
When one uses the rules of reason and logic alone to arrive at truth, this is called (2) ________.	(2) rationalism
Sometimes a sudden insight provides a means of acquiring knowledge. This is referred to as (3) ________.The most refined method of acquiring	(3) intuition
knowledge is by using the (4) ________ (5) ________. The scientific method	(4) scientific (5) method
uses reasoning and intuition but it relies on (6) ________ (7) ________ using (8) ________ to confirm or refute hypotheses.	(6) objective (7) assessment (8) experimentation
Scientific research utilizes two basic research methods to acquire knowledge.	
These are (9) ________ studies and (10) ________ experiments.	(9) observational (10) true
In observational studies the experimenter (11) ________ actively manipulate variables. This method employs	(11) does not

(12) _______ observation, (13) _______	(12) naturalistic
	(13) parameter
estimation, and (14) _______ studies.	(14) correlational-
In many of these studies, the goal is to accurately	
(15) _______ a situation or relationship.	(15) describe
In true experiments the goal is to determine	
whether changes in one variable (16) _______	(16) produce
(17) _______ in another variable. The variable	(17) changes
that the experimenter has control over and	
manipulates is the (18) _______ variable.	(18) independent
The variable that is observed for changes	
is called the	
(19) _______ variable. It is usually not possible	(19) dependent
to collect data on the entire (20) _______ so a	(20) population
subset, called a (21) _______ is studied. It is	(21) sample
crucial that this sample is a	
(22) _______ sample. This helps assure that	(22) random
the laws of (23) _______ apply to the data	(23) probability
and that the sample is (24) _______ of the	(24) representative
population.	

The study of statistics is divided into two areas. Thefirst is (25) _______ (26) _______ which describes and characterizes data.	(25) descriptive (26) statistics
The second area, which allows us to use sample data to infer conclusions about the population is called (27) _______ (28) _______.	(27) inferential (28) statistics
The complete set of individuals, objects, or scores that the investigator is interested in is called the (29) _______.	(29) population
The subset of the population which the experimenter studies is called a (30) _______.	(30) sample
Properties or characteristics of some event, object or person which can take on different values are called (31) _______. Measurements	(31) variables
made on the dependent variable are called (32) _______. Data which	(32) data
have not yet been analyzed or summarized are referred to as (33) _______ data or	(33) raw

(34) _______ scores. A (35) _______ is a measure of a sample characteristic. A (36) _______ is a measure of a population characteristic. A statistic and a parameter are related concepts. A statistic refers to (37) _______ data, while a parameter refers to the entire (38) _______.

(34) original
(35) statistic
(36) parameter
(37) sample
(38) population

EXERCISES

1. What is the difference between rationalism and the scientific method?

2. A social scientist measured the number of years of education of a random group of people and asked them their income. The scientist then found that the longer the education, the higher the income.

   a. What kind of study was this?
   b. Do you think that the scientist can rightfully infer that higher education causes higher income?

3. A scientist takes a random sample of 50 citizens in a city and asks them their age. The scientist then calculates the average age and uses that as an estimate of the average age of all the citizens in the city. What is this process called?

4. Another scientist is interested in parameter estimation. When asked to estimate the average income in a city he went to the Millionaire's Club and asked each person their average income. What was wrong with that process?

5. Consider yourself and describe 3 populations to which you belong?

6. How often are population parameters known?

7. Give three examples of a variable applying to people.

8. In an experiment, a scientist is interested in the effect of a drug on reaction time. He gives various doses of the drug and then measures blood levels of the drug and reaction time in a driving task. What is the independent variable? The dependent variable?

9. If someone asked you the average height of everyone in your family and you actually measured each of the people in your family to calculate the average, have you calculated a statistic or parameter?

10. A student hears a lecture on the harmful effects of drug *X* on intellectual functioning. The student then tells his friend that drug *X* is harmful. What method of inquiry has the student relied upon to learn about drug *X*?

TRUE-FALSE QUESTIONS

T F 1. Intuition is not a valid way of developing research hypotheses.

T F 2. The main difference between rationalism and the scientific method is that the scientific method relies on objective assessment to test the idea.

T F 3. In observational studies the researcher manipulates the experiment and observes the results closely.

T F 4. One task of research is to study a subset of the population and try to make inferences about the whole population.

T F 5. Naturalistic observation is rarely used by scientists because it does not make use of the scientific method.

T F 6. Descriptive statistics are used to make inferences about the population.

T F 7. In a study of caloric intake on weight loss, the number of pounds lost (or gained) is the independent variable.

T F 8. In inferential statistics random sampling is regarded as desirable but not essential.

T F 9. If one measured the height of all the women at a university and took an average of the data, the university women would be a sample of the population of all university women in the world.

T F 10. When data has not yet been analyzed, this is called raw data.

SELF-QUIZ

1. When one analyzes data based on a sample, one calculates a ________.

    a. parameter
    b. variable
    c. constant
    d. statistic

2. Mathematical methods used to draw tentative conclusions about a population based on sample data are referred to as ________.

    a. descriptive statistics
    b. sample statistics
    c. inferential statistics
    d. random sampling
    e. magic

3. The variable that the experimenter manipulates is called the ________.

    a. independent variable
    b. dependent variable
    c. constant variable
    d. experimental variable

4. In acquiring knowledge the method that employs logic, reasoning and objective assessment is referred to as ________.

    a. the method of authority
    b. intuition
    c. rationalism
    d. scientific method

5. From which of the following studies can one most reasonably determine cause and effect?

   a. correlational study
   b. true experiment
   c. naturalistic observation
   d. all the above equally well

6. In inferential statistics the object is usually to generalize from a _______ to a _______.

   a. data; variable
   b. sample; population
   c. population; sample
   d. constant; variable

7. For a list of data the lowest and highest score values are examples of _______ statistics.

   a. inferential
   b. sample
   c. population
   d. descriptive

8. To avoid unknown, systematic factors that may bias the results of an experiment, the experimenter should select a _______ sample from a population and use controlled conditions.

   a. wise
   b. small
   c. random
   d. single

9. Descriptive statistics are helpful in _______ and _______ raw data.

   a. generalizing; inferring
   b. organizing; tenderizing
   c. confusing; confounding
   d. summarizing; characterizing

10. I am certain I will find this course _______.

    a. fun
    b. fatal
    c. interesting
    d. insufficient data to answer question

ANSWERS

Exercises: 1. Both use reason to arrive at knowledge, but in addition, scientific method uses objective assessment to test the validity of an idea. **2a.** correlational study **2b.** No, just because two variables are related, doesn't mean one causes the other. (We'll talk more about this topic in Chapter 7. **3.** Inferential statistics. In this case it is parameter estimation in which the parameter being estimated is the average age of the citizens in the city. **4.** The sample was not random, so the results can only be generalized to the Millionaire's Club. **5.** Any three will do. Some that may apply are college students, males/females, residents of your city, citizens of a country, members of a particular religious group, etc. **6.** Rarely, usually a sample is used to estimate the parameter. **7.** Age, hair color, number of teeth missing, visual acuity, income, religion, race, sex, social security number, etc. **8.** The independent variable is the drug dosage and the dependent variable is reaction time. **9.** population parameter. **10.** authority.

True-False: **1.** F **2.** T **3.** F **4.** T **5.** F **6.** F **7.** F **8.** F **9.** T **10.** T.

Self-Quiz: **1.** d **2.** c **3.** a **4.** d **5.** b **6.** b **7.** d **8.** c **9.** d **10.** Give yourself credit for any answer other than b.

Chapter 2

BASIC MATHEMATICAL AND MEASUREMENT CONCEPTS

CHAPTER OUTLINE

I. Study Hints for the Student

A Review basic algebra but don't be afraid that the mathematics will be too hard.

B. Become very familiar with the notations in the book.

C. Don't fall behind. The material in the book is cumulative and getting behind is a bad idea.

D. Work problems!

II Mathematical Notation

A. Symbols. The symbols *X* (capital letter *X*) and sometimes *Y* will be used as symbols to represent variables measured in the study.

1. For example, X could stand for age, or height, or IQ in any given study.
2. To indicate a specific observation a subscript on X will be used; e.g., X_2 would mean the second observation of the X variable.

B. Summation sign. The summation sign (Σ) is used to indicate the fact that the scores following the summation sign are to be added up. The notations above and below the Σ sign are used to indicate the first and last scores to be summed.

C. Summation rules.

1. The sum of the values of a variable plus a constant is equal to the sum of the values of the variable plus N times the constant. In equation form

$$\sum_{i=1}^{N}(X_i + a) = \sum_{i=1}^{N} X_i + Na$$

2. The sum of the values of a variable minus a constant is equal to the sum of the variable minus N times the constant. In equation form

$$\sum_{i=1}^{N}(X_i - a) = \sum_{i=1}^{N} X_i - Na$$

3. The sum of a constant times the values of a variable is equal to the constant times the sum of the values of the variable. In equation form

$$\sum_{i=1}^{N} aX_i = a\sum_{i=1}^{N} X_i$$

4. The sum of a constant divided into the values of a variable is equal to the constant divided into the sum of the values of the variable. In equation form

$$\sum_{i=1}^{N} (X_i / a) = (\sum_{i=1}^{N} X_i) / a$$

III. Measurement Scales.

A. Attributes. All measurement scales have one or more of the following three attributes.

1. Magnitude.
2. Equal intervals between adjacent units.
3. Absolute zero point.

B. Nominal scales. The nominal scale is the lowest level of measurement. It is more qualitative than quantitative. Nominal scales are comprised of elements that have been classified as belonging to a certain category. For example, whether someone's sex is male or female. Can only determine whether $A = B$ or $A \neq B$.

C. Ordinal scales. Ordinal scales possess a relatively low level of the property of magnitude. The rank order of people according to height is an example of an ordinal scale. One does not know how much taller the first rank person is over the second rank person. Can determine whether $A > B$, $A = B$ or $A < B$.

D. Interval scales. This scale possesses equal intervals, magnitude, but no absolute zero point. An example is temperature measured in degrees Celsius. What is called zero is actually the freezing point of water, not absolute zero. Can do same determinations as ordinal scale, plus can determine if $A - B = C - D$, $A - B > C - D$, or $A - B < C - D$.

E. Ratio scales. These scales have the most useful characteristics since they possess attributes of magnitude, equal intervals, and an absolute zero point. All mathematical operations can be performed on ratio scales. Examples include height measured in centimeters, reaction time measured in milliseconds.

IV. Additional Points Concerning Variables

A. Continuous variables. This type can be identified by the fact that they can theoretically take on an infinite number of values between adjacent units on the scale. Examples include length, time and weight. For example, there are an infinite number of possible values between 1.0 and 1.1 centimeters.

B. Discrete variables. In this case there are no possible values between adjacent units on the measuring scale. For example, the number of people in a room has to be measured in discrete units. One cannot reasonably have 6 1/2 people in a room.

C. Continuous variables. All measurements on a continuous variable are approximate. They are limited by the accuracy of the measurement instrument. When a measurement is taken, one is actually specifying a range of values and calling it a specific value. The real limits of a continuous variable are those values that are above and below the recorded value by 1/2 of the smallest measuring unit of the scale (e.g., the real limits of 100°C are 99.5° C and 100.5° C, when using a thermometer with accuracy to the nearest degree).

D. Significant figures. The number of decimal places in statistics is established by tradition. The advent of calculators has made carrying out laborious calculations much less cumbersome. Because solutions to problems often involve a large number of intermediate steps, small rounding inaccuracies can become large errors. Therefore, the more decimals carried in intermediate calculations, the more accurate is the final answer. It is standard practice to carry to one or more decimal places in intermediate calculations than you report in the final answer.

E. Rounding. If the remainder beyond the last digit is greater than 1/2 add one to the last digit. If the remainder is less than 1/2 leave the last digit the same. If the remainder is equal to 1/2 add one to the last digit if it is an odd number, but if it is even, leave it as it is.

CONCEPT REVIEW

In statistics we often let symbols stand for variables.

Generally, we will let the letter (1) _______ stand for a variable such as height, IQ or reaction time.

(1) *X*

We will also sometimes use the letter *Y*.

(2) subscripts

We will use (2) _______ on X to indicate a specific case. For example, one would designate the third value of IQ in a set of scores as (3) _______. One would — (3) X_3

indicate the nth score as (4) _______. One — (4) X_n

very common operation is to sum all the scores.

To indicate this operation, we use the symbol — (5) Σ

(5) _______, which is the Greek capital letter — (6) sigma

(6) _______. Summing a set of scores simply

means to (7) _______. Algebraically, we can write a — (7) add them

mathematical sentence which can easily be translated

into English. The symbols $\sum_{i=1}^{N} X_i$ can be translated

as meaning (8) _______ of the (9) _______ — (8) sum

variable from i = (10) _______ to (11) _______. — (9) X

(10) 1

The notations above and below the summation — (11) N

sign designate which (12) _______ to include in — (12) scores

the summation. The term (13) _______ the summation — (13) below

sign tells us the first score in the summation. The

term above the Σ designates the (14) _______

score. If the summation is over all — (14) last

the scores (from 1 to N) we can abbreviate

the mathematical sentence to (15) _______. If	
one wanted to indicate summing the 3rd, 4th,	(15) ΣX
and 5th scores, one could show this mathematically	
as (16) _______. The expression for	(16) $\sum_{i=3}^{N} X_i$
summing all the squared *X* scores is (17) _______.	(17) ΣX^2
The notation for summing all the *X* scores and	
squaring the final sum is (18) _______.	(18) $(\Sigma X)^2$
There are four types of measurement scales.	
Measurements which classify objects into groups	
with names are called (19) _______ scales.	
This is primarily a measurement which	(19) nominal
names objects. Nominal scales are generally	
(20) _______ not quantitative.	(20) qualitative
The (21) _______ scale is the next higher	(21) ordinal
level of measurement. Ordinal scales possess a	
low level of the property of (22) _______. A	(22) magnitude
common application of ordinal scales is to	
(23) _______ order objects. The rank order tells	(23) rank

which object has (24) _______ or (25) _______ of a given attribute but not how much more or less.

(24) more
(25) less

Thus, it can be said that ordinal scales do not have (26) _______ intervals between adjacent units.

(26) equal

A scale which possess both properties of magnitude and equal intervals between adjacent units is the (27) _______ scale. The difference between an interval scale and a (28) _______ scale is that in addition, a ratio scale possesses an (29) _______ zero point. The Celsius temperature scale would be an example of an (30) _______ scale because the zero point on that scale is not at absolute zero. With interval scales, we can tell whether two objects possess more, less or the same amount of a given attribute as well as (31) _______ of the attribute. Addition, subtraction, multiplication, division and doing ratios are

(27) interval
(28) ratio
(29) absolute
(30) interval
(31) how much more or less

permitted with (32) _______ scales	(32) ratio
Earlier we defined a variable as a property	
or characteristic of something which can take	(33) value
more than one (33) _______. Variables can be	
either continuous or discrete. A continuous variable	
can have an (34) _______ number of values between	(34) infinite
adjacent units. Discrete variables have (35) _______	(35) no
possible values between adjacent units. Length	
measured in centimeters is an example of	
a (36) _______ variable. The number	(36) continuous
of siblings in a specific family is an example	(37) discrete
of a (37) _______ variable. Because of their	
nature, measures on continuous variables are	
(38) _______. We aren't able to measure the	(38) approx-imate
exact value for a continuous variable but we	
can specify the (39) _______ limits for a value.	
The real limits are those	(39) real
values which are above and below the recorded	

value by one-half of the (40) _______ measuring unit

of the scale. If we were measuring distance (40) smallest

to the nearest mile and observed that a trip was

16 miles long, the lower real limit is (41) _______ (41) 15.5

miles and the upper real limit is (42) _______ miles. (42) 16.5

In rounding answers in statistics we use the rule:

If the remainder beyond the last digit is greater than

1/2, (43) _______ one to the last digit. If the remainder (43) add

is less than 1/2, (44) _______ the last digit as it is. (44) leave

If the remainder is equal to 1/2, add one to the last

digit if it is an (45) _______ number, but (45) odd

if it is (46) _______, leave it as it is. Thus, (46) even

rounding 3.14159 to four decimals gives (47) _______. (47) 3.1416

Rounding 3.5 to a whole number gives (48) _______. (48) 4

Rounding 3.650 to one decimal gives

(49) _______. Rounding 2.5001 to the nearest (49) 3.6

whole number gives (50) _______. (50) 3

EXERCISES

1. Consider the following sample scores for the variable weight:

$$X_1 = 145, X_2 = 160, X_3 = 110, X_4 = 130, X_5 = 137, X_6 = 172, \text{ and } X_7 = 150$$

   a. What is the value for $\sum X$?
   b. What is the value for $\sum_{i=3}^{6} X_i$?
   c. What is the value for $\sum X^2$?
   d. What is the value for $\left(\sum X\right)^2$?
   e. What is the value for $\sum (X+4)$?
   f. What is the value for $\sum X - 140$?
   g. What is the value for $\sum (X - 140)$?

2. Round the following values to one decimal place.

   a. 25.15
   b. 25.25
   c. 25.25001
   d. 25.14999
   e. 25.26

3. State the real limits for the following values of a continuous variable.

   a. 100 (smallest unit of measurement is 1)
   b. 1.35 (smallest unit of measurement is 0.01)
   c. 29.1 (smallest unit of measurement is 0.1)

4. Indicate whether the following variables are discrete or continuous.

   a. The age of an experimental subject.
   b. The number of ducks on a pond.
   c. The reaction time of a subject on a driving task.
   d. A rating of leadership on a 3-point scale.

5. Identify which type of measurement scale is involved for the following:

   a. The sex of a child.
   b. The religion of an individual
   c. The rank of a student in an academic class.
   d. The attitude score of a subject on a prejudice inventory.
   e. The time required to complete a task.
   f. The rating of a task as either "easy," "mildly difficult," or "difficult."

6. In an experiment measuring the number of aggressive acts of six children, the following scores were obtained.

Subject	Number of Aggressive Acts
1	15
2	25
3	5
4	18
5	14
6	22

   a. If X represents the variable of “Number of Aggressive Acts”, assign each of the scores its appropriate X symbol.
   b. Compute ΣX for these data.

7. Given the following sample scores for the variable length (cm):

$$X_1 = 22,\ X_2 = 35,\ X_3 = 32,\ X_4 = 43,\ X_5 = 28$$

   a. What is the value for $\sum(X+4)$?
   b. What is the value for $\sum X - 15$?
   c. What is the value for $\sum(X - 15)$?

8. Using the scores in Problem 7:

   a. What is the value for $\sum\left(\frac{X}{3}\right)$?
   b. What is the value for $\sum 5X$?

9. Using the scores in Problem 7:

   a. What is the value for $(\sum X)^2$?

   b. What is the value for $\sum X^2$?

10. Round the following to two decimal place accuracy.

   a. 75.0338
   b. 75.0372
   c. 75.0350
   d. 75.0450
   e. 75.045000001

TRUE-FALSE QUESTIONS

T F 1. All scales possess magnitude, equal intervals between adjacent units, and an absolute zero point.

T F 2. Nominal scales can be used either qualitatively or quantitatively.

T F 3. With an ordinal scale one cannot be certain that the magnitude of the distance between any two adjacent points is the same.

T F 4. With the exception of division, one can perform all mathematical operations on a ratio scale.

T F 5. The average number of children in a classroom is an example of a discrete variable.

T F 6. When a weight is measured to 1/1000th of a gram, that measure is absolutely accurate.

T F 7. If the quantity $\Sigma X = 400.3$ for N observations, then the quantity ΣX will equal 40.03 if each of the original observations is multiplied by 0.1.

T F 8. One generally has to specify the real limits for discrete variables since they cannot be measured accurately.

T F 9. The symbol Σ means square the following numbers and sum them.

T F 10. Rounding 55.55 to the nearest whole number gives 55.

SELF-QUIZ

Refer to the following set of numbers to answer questions 1 - 6:

$$X_1 = 2, X_2 = 4, X_3 = 6, X_4 = 10$$

1. What is the value for ΣX?

   a. 12
   b. 156
   c. 480
   d. 22

2. What is the value of ΣX^2?

   a. 156
   b. 22
   c. 480
   d. 37

3. What is the value of X_4^2?

   a. 4
   b. 6
   c. 100
   d. 10

4. What is the value of $(\Sigma X)^2$?

   a. 480
   b. 484
   c. 156
   d. 44

5. What is the value of N?

   a. 2
   b. 4
   c. 6
   d. 10

6. What is the value of $(\Sigma X)/N$?

   a. 5
   b. 4
   c. 6
   d. 5.5

7. Classifying subjects on the basis of sex is an example of using what kind of scale?

   a. nominal
   b. ordinal
   c. interval
   d. ratio
   e. bathroom

8. Number of bar presses is an example of a(n) _______ variable.

   a. discrete
   b. continuous
   c. nominal
   d. ordinal

9. Using an ordinal scale to assess leadership, which of the following statements is appropriate?

   a. *A* has twice as much leadership ability as *B*
   b. *X* has no leadership ability

c. *Y* has the most leadership ability
d. all of the above

10. The number of legs on a centipede is an example of a _______ scale.

    a. nominal
    b. ordinal
    c. ratio
    d. continuous

11. What are the real limits of the observation of 6 seconds (measured to the nearest second)?

    a. 5.5-6.5
    b. 5.0-7.0
    c. 5.9-6.1
    d. 6.0-6.5

12. What is 17.295 rounded to one decimal place?

    a. 17.1
    b. 17.0
    c. 17.2
    d. 17.3

13. What is the value of 0.05 rounded to one decimal place?

    a. 0.0
    b. 0.1
    c. 0.2
    d. 0.5

14. The symbol "Σ" means:

    a. add the scores
    b. summarize the data
    c. square the value
    d. multiply the scores

The following problems are for your own use in evaluating your skills at elementary algebra. If you do not get all the problems correct you should probably review your algebra.

15. Where $3X = 9$, what is the value of X?

    a. 3
    b. 6
    c. 9
    d. 12

16. For $X + Y = Z$, X equals _______.

    a. $Y + Z$
    b. $Z - Y$
    c. Z/Y
    d. Y/Z

17. $1/X + 2/X$ equals _______.

    a. $2/X$
    b. $3/2X$
    c. $3/X$
    d. $2/X^2$

18. What is $(4 - 2)(3 \bullet 4)/(6/3)$?

    a. 24
    b. 1.3
    c. 12
    d. 6

19. $6 + 4 \times 3 - 1$ simplified is _______.

    a. 29
    b. 48
    c. 71
    d. 17

20. $X = Y/Z$ can be expressed as _______.

    a. $Y = (Z)(X)$
    b. $X = Z/Y$
    c. $Y = X/Z$
    d. $Z = X + Y$

21. 2^4 equals _______.

    a. 4
    b. 32
    c. 8
    d. 16

22. $\sqrt{81}$ equals _______.

    a. ± 3
    b. ± 81
    c. ± 9
    d. ± 27

23. $X(Z + Y)$ equals _______.

    a. $XZ + Y$
    b. $ZX + YX$
    c. $(X)(Y)(Z)$
    d. $(Z + Y)/X$

24. 1/2 + 1/4 equals _______.

    a. 1/6
    b. 1/8
    c. 2/8
    d. 3/4

25. X^6/X^2 equals _______.

    a. X^8
    b. X^4
    c. X^2
    d. X^3

ANSWERS

Exercises: 1a. 1004 **1b.** 549 **1c.** 146,478 **1d.** 1,008,016 **1e.** 1032 **1f.** 864 **1g.** 24 4 **2a.** 25.2 **2b.** 25.2 **2c.** 25.3 **2d.** 25.1 **2e.** 25.3 **3a.** 99.5 - 100.5 **3b.** 1.345 - 1.355 **3c.** 29.05 - 29.15 **4a.** continuous **4b.** discrete **4c.** continuous **4d.** discrete **5a.** nominal **5b.** nominal **5c.** ordinal **5d.** interval (assuming equal interval) or ordinal (if equal interval assumption is unreasonable) **5e.** ratio **5f.** ordinal **6a.** $X_1 = 15$, $X_2 = 25$, $X_3 = 5$, $X_4 = 18$, $X_5 = 14$, $X_6 = 22$ **6b.** 99. **7a.** 180 **7b.** 145 **7c.** 85 **8a.** 53.33 **8b.** 800 **9a.** 25,600 **9b.** 5366 **10a.** 75.03 **10b.** 75.04 **10c.** 75.04 **10d.** 75.04 **10e.** 75.05.

True-False: **1.** F **2.** F **3.** T **4.** F **5.** F **6.** F **7.** T **8.** F **9.** F **10.** F.

Self-Quiz: **1.** d **2.** a **3.** c **4.** b **5.** b **6.** d **7.** a **8.** a **9.** c **10.** c **11.** a **12.** d **13.** a **14.** a **15.** a **16.** b **17.** c **18.** c **19.** d **20.** a **21.** d **22.** c **23.** b **24.** d **25.** b.

Chapter 3

FREQUENCY DISTRIBUTIONS

CHAPTER OUTLINE

I. Constructing Frequency Distributions

A. Frequency distribution. A list of rank ordered scores and their frequency of occurrence in tabular form is called a frequency distribution. When the data is rank ordered, it is easier to understand.

B. Grouping data. Individual scores are often grouped together into class intervals of equal width to allow one to visualize the shape of a distribution and its central tendency. This is called a frequency distribution of grouped scores.

1. Distributions are usually divided into 10 to 20 intervals.
2. If the intervals are too wide, too much information is lost; if too narrow, there are too many zero frequency intervals which distort the shape of the distribution.

C. Constructing a frequency distribution of grouped scores.

1. Find the range of scores.

2. Determine the width of each class interval (i).

3. List the limits of each class interval, placing the interval containing the lowest score at the bottom.
4. Tally the raw scores into the appropriate class intervals.
5. Add the tallies to obtain the interval frequency.

II. Desired Frequency Distributions. It is often desirable to express data in forms other than the basic frequency distribution.

A. <u>Relative frequency distribution</u>. This indicates the proportion of the total number of scores that occurred in each interval.

B. <u>Cumulative frequency distribution</u>. This indicates the number of scores which fell below the upper real limit of each interval.

C. <u>Cumulative percentage distribution</u>. This indicates the percentage of scores which fell below the upper real limit of each interval. To convert cumulative frequencies to cumulative percentages one can apply the formula:

$$\textbf{cum \% = (cum } \boldsymbol{f}\textbf{)/}\boldsymbol{N} \times \textbf{100}$$

III. Measures of Relative Standing. These are used to compare performances of an individual to that of a reference group.

A. <u>Percentile point</u>.

1. Percentile point is the value on the measurement scale below which a specified percentage of scores in the distribution fall.
2. Formula.

$$\textbf{Percentile Point} = X_L + (i/f_i)(\text{cum } f_P - \text{cum } f_L)$$

B. Percentile rank.

1. Percentile rank of a score is the percentage of scores lower than the score in question.

2. Formula:

$$\textbf{Percentile rank} = \frac{\text{cum } f_L + (f_i/i)(X - X_L)}{N} \times 100$$

IV. Graphing Techniques

A. Graph characteristics.

1. A graph has two axes, the vertical is called the ordinate and the horizontal is called the abscissa.
2. Scores are shown on the abscissa; frequency is shown on the ordinate.
3. Graphs are generally 3/4 as high as they are long.

B. Types of graphs for frequency distributions.

1. Bar graphs are generally used for nominal or ordinal data. The abscissa shows categories and the ordinate frequency.

2. Histograms are generally used for interval or ratio data. The abscissa shows the class interval for the data and the ordinate shows frequency. Vertical bars must touch each other.
3. Frequency polygons are similar to histograms, except that instead of using bars, a point is plotted over the midpoint of each interval at a height corresponding to the frequency of the interval. In addition, the line is extended to meet the horizontal axis at the midpoint of the two end-adjacent intervals.
4. Cumulative percentage polygons show the percentage of scores that fall below the upper real limit of the interval on the ordinate. The abscissa plots the upper limit of each class interval.

C. Shapes of frequency curves.

1. Symmetrical. Symmetrical distributions are those which when folded in half have the two sides of the curve correspond perfectly.
2. Skewed. If a curve is not symmetrical, it is skewed.

    a. Positive skew refers to curves where most of the scores occur at the lower values of the abscissa and tails off to the higher end.
    b. Negative skew refers to curves where most of the scores occur at the higher values and the curve tails toward the lower end of the abscissa.

3. Miscellaneous shapes. Curves are sometimes described in terms of their shape. Examples include U-shaped distributions, J-shaped distributions, rectangular, and bell-shaped.

D. Exploratory Data Analysis.

1. Uses easy-to-construct diagrams to describe data.
2. Stem and leaf diagrams.

    a. Alternative to histogram, but unlike the histogram, preserves the original scores.
    b. Each score is represented by a stem and leaf. Stems are placed vertically down the page and leafs are placed to the right of the stems, horizontally across the page. Often, a vertical line is used to separate the stems from the leafs.

CONCEPT REVIEW

An unordered list of many scores is difficult to understand.

To report large amounts of data in an informative way, a table is often presented. A (1) _______ presents the score values and their frequency of occurrence.

(1) frequency distribution

The scores are listed in (2) _______ order, generally with the (3) _______ score at the bottom of the table. To avoid having many values with frequencies of zero, the individual scores are usually (4) _______ into (5) _______ intervals of equal width.

(2) rank
(3) lowest
(4) grouped
(5) class

This is called a frequency distribution of (6) ______.	(6) grouped scores
The (7) ______ the interval the more information	(7) wider
is lost. If the interval is too (8) ______ then too	(8) narrow
many zero frequencies occur. Generally	
the distribution is divided into from (9) ______ to	(9) 10
(10) ______ intervals. To construct a	(10) 20
frequency distribution, one employs the following steps:	
1. Find the (11) ______ of the scores. To	(11) range
determine the range, one subtracts the (12) ______	(12) lowest
value from the (13) ______ value in the distribution.	(13) highest
2. Determine the (14) ______ of the class interval.	(14) width
The width of the class interval is given by the symbol	
(15) ______. The width of the class interval is	(15) *i*
determined by dividing the (16) ______ by the	(16) range
(17) ______ of class intervals. If there is a	(17) number
decimal remainder, round to the	
(18) ______ number of digits as the raw scores.	(18) same
3. When listing the intervals, begin	

with the (19) _______interval and proceed to the highest. First, one needs to determine the (20) _______ of the lowest interval. Two criteria exist in determining the values of the lowest intervals. The interval must begin with a value (21) _______ or (22) _______the lowest score. Also, the lower limit should be evenly (23) _______ by *i*.

(19) lowest

(20) lower limit

(21) equal to

(22) less than

(23) divisible

4. Next, the raw scores are (24) _______ into the appropriate class intervals. This simply entails entering a tally mark next to the interval containing the raw score.

(24) tallied

5. Count the tally marks and convert them into (25) _______.

(25) frequencies

In addition to simply counting up frequencies, it is often useful to summarize them further. A (26) _______ frequency distribution indicates the proportion of the total number of scores that occurred in each (27) _______. A

(26) relative

(27) interval

(28) _______ frequency distribution indicates the number of scores that fell below the (29) _______ real limit of each interval. A cumulative (30) _______ distribution indicates the percentage of scores which fell below the upper real limit of each interval.	(28) cumulative (29) upper (30) percent age
To convert a frequency distribution into a relative frequency distribution, the (31) _______ for each interval is (32) _______ by the total number of scores. The cumulative frequency for each interval is found by (33) _______ the frequency of that interval to the sum of the frequencies of all the class intervals below.	(31) frequency (32) divided (33) adding
The cumulative percentage for each interval is found simply by converting cumulative (34) _______ to cumulative (35) _______. The equation for this is:	(34) frequencies (35) percent-ages

$$\text{cum } \% = (\text{cum } f)/N \times 100$$

Cum % is the (36) _______. N equals the	(36) cumulative

(37) _______ number of scores. Cum *f* equals	(37) percentage total
the cumulative (38) _______ of the class interval	(38) frequency
of interest.	
If there were 180 scores altogether and the	
cumulative frequency of a given class interval	
equaled 40, the cumulative percentage would	
be (39) _______. This would mean that	(39) cum % = (40/180)× 100 = 22.22%
(40) _______ percent of the scores fell below the	(40) 22.22
upper real limit of the interval in question.	
Percentages are measures of relative standing.	
They are used to (41) _______ the performance	(41) compare
of an individual to that of a reference group. A	
(42) _______ is the value on the measurement scale	(42) percentile or percen tile point
below which point a specified percentage of the	
scores in the distribution falls. The specified	

percentage is called the (43) _______ of the value. For example, the 38th percentile point is the value below which (44) _______ percent of the scores in the distribution fall. This value has a percentile (45) _______ of (46) _______.

(43) percentile rank
(44) 38
(45) rank
(46) 38

There is a formula to compute the percentile point. The formula is:

Percentile Point = $X_L + (i/f_i)(\text{cum } f_P - \text{cum } f_L)$

In order to avoid difficulties in understanding and applying the equation, one needs to be certain that one understands what the symbols mean. In this equation:

X_L = the value of the (47) _______ real limit of the interval (48) _______ the percentile point. cum f_P = the frequency of scores (49) _______ the percentile point.

(47) lower
(48) containing
(49) below

cum f_L = the frequency of scores (50) _______ the

(50) below

(51) _______ real limit of the interval containing the percentile point. f_i = the frequency of	(51) lower
the interval (52) _______ the percentile	(52) containing point.
i = the (53) _______ of the (54) _______.	(53) width (54) interval
Now is a good time to refresh the memory on the rules of algebra as they apply to simplifying this equation. The first step is to simplify the	
terms within (55) _______. The second step	(55) parentheses
is to (56) _______ the terms in parentheses together. Finally, one adds this product to	(56) multiply
(57) _______ to obtain the	(57) X_L
(58) _______.	(58) percentile point
It is often useful to know the percentile rank of a raw score. The percentile rank of	
a score is the (59) _______ of scores	(59) percentage
(60) _______ than the score in question.	(60) lower
This process is the (61) _______ situation than	(61) opposite
the process of calculating the (62) _______.	(62) percentile point

In this case we are given the (63) _______ (63) raw score

and must calculate the (64) _______ of scores (64) percent age

below it. In the percentile point calculation we

are given the (65) _______ and must calculate the (65) percentile

(66) _______. The formula for the (66) scale value

percentile rank is as follows:

$$\text{Percentile rank} = \frac{\text{cum}\, f_L + (f_i/i)(X - X_L)}{N} \times 100$$

where

cum f_L = the frequency of scores (67) _______ (67) below

the (68) _______ real limit of the (68) lower

interval containing the score X.

X = the raw score whose (69) _______ is (69) percentile rank

being determined. X_L = the (70) _______ of the (70) value

lower real limit of the

interval containing (71) _______. (71) X

i = (72) _______. (72) interval width

f_i = the (73) _______ of the interval containing X. (73) frequency

N = the (74) _______ number of scores. (74) total

The same rules of algebra apply to simplifying this equation as did for the previous one.

Graphs are often used to show frequency distributions. They present (75) _______ new information from the tables discussed above, but a picture can sometimes convey information more clearly. Graphs have two axes. The vertical axis is called the (76) _______. The horizontal axis is called the (77) _______. In graphing frequency distributions, the score (78) _______ are plotted along the (79) _______. It is customary to have the height of the graph be about (80) _______ of the length.

(75) no
(76) ordinate
(77) abscissa
(78) frequencies
(79) ordinate
(80) 3/4

One common way of plotting nominal or ordinal data is using the (81) _______ graph. A bar is drawn for each

(81) bar

(82) _______ and the height of the bar represents (82) category

(83) _______. In the case of (84) _______ data (83) frequency (84) nominal

the order of arrangement of the groups along

the horizontal axis does not matter. The bars do

not touch each other to emphasize the lack of a

(85) _______ relationship between the categories. (85) quantitative

A (86) _______ can be used to represent (86) histogram

frequency distributions of interval or (87) _______

scales. In a histogram a bar is drawn for each (87) ratio

(88) _______. Usually the (89) _______ of the (88) class interval

class intervals are plotted along the (90) _______. (89) midpoints (90) abscissa

The height of the bar corresponds to the

(91) _______ of the class interval. Since the (91) frequency

data are (92) _______, the bars touch each other. (92) continuous

A frequency (93) _______ can also be used (93) polygon

to represent interval or ratio data. Instead of

using bars, a point is plotted

over the (94) _______ of each class interval (94) midpoint

at a height corresponding to the (95) _______ of

the interval. At the ends of the distribution,	(95) frequency
the lines joining the points are extended to	
meet the (96) _______ axis to form a polygon.	(96) horizontal
The cumulative percentage of a distribution	
can easily be calculated and, therefore, graphed.	
When graphed this is called a (97) _______ curve.	(97) cumulative percentage
In this graph the vertical axis is scaled in (98) _____	(98) cumulative
units. On the horizontal axis the upper real	
limit of the class interval is plotted. One can	percentage
read both (99) _______ and(100) _______ from a	(99) percentiles
cumulative percentage curve. The cumulative	(100) percentile rank
percentage curve is also called an (101) _______,	(101) ogive
meaning S-shape.	
Frequency curves can be described in terms	
of their shapes. They are generally classified as to	
whether they are (102) _______ or (103) _______.	(102) symmetrical (103) skewed
A curve is (104) _______	(104) symmetrical

if when folded in half the two sides coincide.

Distributions which are not symmetrical are

(105) _______. A positively skewed distribution (105) skewed

has most of the scores at the (106) _______ (106) lower

values of the distribution and tails off to the

(107) _______ values. A negatively skewed (107) higher

distribution has most of the scores at the (108) ______ (108) higher

values of the distribution and tails off toward

the (109) _______ end of the horizontal axis. (109) lower

Frequency distributions are often referred to

according to their (110) _______; e.g., bell-shaped, (110) shape

U-shaped, J-shaped, or rectangular, to just

name a few.

EXERCISES

1. The first step in making a frequency distribution is to rank order the scores. Generally this is done by ranking scores from lowest to highest. Rank order the following scores:

30, 20, 21, 25, 5, 18, 18, 16, 10

2. What is the range of the distribution in problem 1?

3. Assume the following are scores in a 100-point achievement test:

58	70	68	59	62
47	90	57	45	68
93	83	45	48	55
80	61	80	70	51
35	58	36	63	71
60	41	65	84	42
40	87	75	81	80
61	72	76	73	52
63	71	94	66	
68	69	60	61	

a. What is the range of the distribution?
b. Since there is no definitive rule for determining how many class intervals to divide data into, one generally plays with the data or is told how many intervals to use. If you were told to group the above data into approximately 12 intervals of equal width, how wide would the class intervals be? Using the value just calculated for i, what would be the lower apparent limit of the lowest class interval?
c. Using 12 intervals construct a table that shows the frequency distribution of grouped scores, the corresponding cumulative frequency distribution, and the corresponding cumulative percentage distribution.
d. Plot the cumulative percentage curve for this data.

4. You have just been hired as a statistician by the federal government. Your first assignment is to prepare a graphic representation of the number of types of animals in a certain national forest. You are told there are 525 birds, 100

beavers, 150 bears, 300 deer, 250 elk, 400 squirrels, 600 rabbits, and 200 geese. Prepare a graph to show these data.

5. Sketch a cumulative frequency distribution for the following frequency distribution.

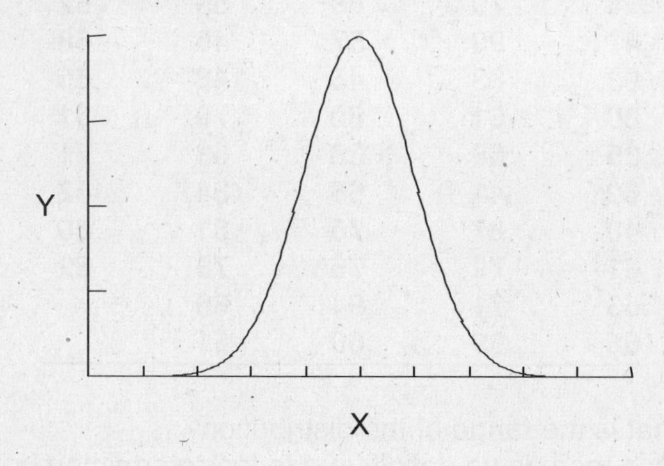

6. Consider the frequency distribution shown in the table below.

class interval	f	cum f	cum %
45-49	2	128	100.00
40-44	6	126	98.44
35-39	15	120	93.75
30-34	24	105	82.03
25-29	30	81	63.28
20-24	22	51	39.84
15-19	16	29	22.66
10-14	8	13	10.16
5-9	4	5	3.91
0-4	1	1	0.78

a. What is the percentile point for the 25th percentile (P_{25})?

b. For the 50th percentile (P_{50})?
c. For the 75th percentile (P_{75})?

7. Again consider the data from problem 6.

a. What percentile rank would a score of 24 have?
b. What percentile rank would a score of 32 have?
c. What percentile rank would a score of 35 have?

8. Plot a cumulative percentage distribution for the data in problem 6.

9. What is wrong with the following frequency distribution?

class interval	f
80-99	51
60-79	59
40-59	112
20-39	49
0-19	50

10. Assume the following are weights of a sample of junior high school students.

98	133	108	106	104
124	110	137	119	110
120	130	112	129	126
112	111	120	103	124
126	119	130	115	118
102	121	125	118	122
116	139	134	123	114
106	112	108	116	120

a. Construct a frequency distribution for the data with approximately 10 intervals.
b. What is the value for P_{50}?

c. What is the value for P_{90}?

d. What is the percentile rank of a weight of 133 pounds?
e. What is the percentile rank of 105 pounds.

11. Construct a stem and leaf diagram for the frequency distribution shown in problem 10.

TRUE-FALSE QUESTIONS

T F 1. When constructing frequency distributions there must be 12 class intervals.

T F 2. In a frequency distribution the more intervals the better, regardless of whether some intervals have zero frequency.

T F 3. One reason for constructing frequency distributions is to be able to visualize the shape of the distribution.

T F 4. When constructing the frequency distribution it is customary to show the real limits of the class intervals in the table.

T F 5. A cumulative frequency distribution indicates the number of scores which fell below the upper real limit of each interval.

T F 6. A relative frequency distribution indicates the total number of scores which occurred in each interval.

T F 7. The vertical axis, i.e., the *Y* axis of a graph, is called the abscissa.

T F 8. Bar graphs are generally used for nominal or ordinal data and histograms are generally used for interval or ratio data.

T F 9. A "U" shaped distribution is an example of a symmetrical distribution.

T F 10. In a cumulative percentage curve, percentage is shown on the abscissa.

T F 11. To determine the width of the class interval, i, divide the range by the number of class intervals.

T F 12. The lower limit in the lowest class interval should equal *i*.

T F 13. Stem and leaf diagrams are like histograms, except that they lose the original scores.

SELF-QUIZ

1. The purpose of a frequency distribution is to _______.

   a. present scores and their frequency of occurrence
   b. present data in a more meaningful way than presenting just the raw scores
   c. provide more information than a graph
   d. all of the above
   e. a and b

2. The true limits of 7.0 are _______.

   a. 6.5-7.5
   b. 6.0-8.0

c. 7.0-7.1
d. 6.95-7.05

3. When individual scores are combined into groups, _______.

a. information is lost
b. data is added
c. a meaningful visual display can result depending on the interval width
d. a and c

4. The range of a set of scores with a maximum value of 92 and a minimum value of 26 is _______.

a. 65
b. 66
c. 67
d. 92

5. If the range of a distribution were 89 and the data were reported as whole numbers, what would the width of the class interval be if one chose to group the distribution into approximately 14 class intervals?

a. 14
b. 89
c. 5
d. 6
e. 7

6. If $i = 7$, and the minimum value of a distribution of scores was 8, what would the lowest class interval be?

a. 0-7
b. 7-14
c. 8-16
d. 8-15

e. 7-13

7. What indicates the proportion of the total number of scores which occurred in each interval?

   a. relative frequency distribution
   b. cumulative frequency distribution
   c. cumulative percentage distribution
   d. none of the above

8. What indicates the number of scores that fell below the upper real limit of each interval?

   a. relative frequency distribution
   b. cumulative frequency distribution
   c. cumulative percentage distribution
   d. none of the above

9. In graphing frequency distributions, _______ is usually plotted on the abscissa.

   a. frequency
   b. class width
   c. the score value
   d. interval width

10. When constructing bar graphs, the bars do not touch each other because _______.

    a. it looks nicer
    b. it emphasizes the lack of quantitative relationship between the categories
    c. it is traditional
    d. none of the above

11. In a frequency polygon the points are plotted over _______ at a height corresponding to the frequency of the interval.

a. the midpoint of each interval
b. the lower real limit
c. the upper real limit
d. none of the above

12. Which of the following is (are) not a symmetrical distribution?

a. a bell-shaped curve
b. a J-shaped curve
c. a rectangular curve
d. an inverted U-shaped curve
e. a, b and d

13. A distribution which has a predominance of scores at the lower values of the distribution and which tails off at the higher end is _______.

a. positively skewed
b. negatively skewed
c. normally distributed
d. symmetrical

For questions 14 - 22 refer to the following table.

class interval	f	cum f	cum %	relative %
90-99	2	139	100.00	1.44
80-89	7	137	98.56	5.04
70-79	15	130	93.53	10.79
60-69	21	115	82.73	15.11
50-59	37	94	67.63	26.62
40-49	26	57	41.01	18.71
30-39	19	31	22.30	13.67
20-29	6	12	8.63	4.32
10-19	4	6	4.32	2.88
0-9	2	2	1.44	1.44

14. How many occurrences are there for the interval 60-69?

15. How many occurrences fall below the upper real limit of the interval 70-79?

16. What percentage of cases fall within the class interval containing the most cases?

17. N equals ________.

18. What is the cumulative percentage below the lower real limit of the interval 90-99?

19. What is the value of i?

20. The cumulative frequency of 115 indicates that 115 scores fall below ________.

    a. 115
    b. 60
    c. 69

d. 69.5
e. 60.5
f. 64.5

21. The percentile rank of a score of 41 equals _______.

22. The 50th percentile point equals _______.

ANSWERS

Exercises: 1. 5, 10, 16, 18, 18, 20, 21, 25, 30 **2.** 25 **3a.** 59 **3b.** 5, 35 **3c.**

class interval	f	cum f	cum %
90-94	3	48	100.00
85-89	1	45	93.75
80-84	6	44	91.67
75-79	2	38	79.17
70-74	6	36	75.00
65-69	6	30	62.50
60-64	8	24	50.00
55-59	5	16	33.33
50-54	2	11	22.92
45-49	4	9	18.75
40-44	3	5	10.42
35-39	2	2	4.17

4. You should have drawn a bar graph since we are showing nominal data on the *X* axis **5.** You should have drawn an ogive **6a.** 20.18 **6b.** 26.67 **6c.** 32.62 **7a.** 38.12 **7b.** 72.66 **7c.** 83.20 **9.** class intervals are too wide **10a.**

class interval	f
136-139	2
132-135	2
128-131	3
124-127	5
120-123	6
116-119	6
112-115	5
108-111	5
104-107	3
100-103	2
96-99	1

10b. P_{50} is 118.17 **10c.** 131.5 **10d.** 91.88 **10e.** 10.31.

11.

9	**8**
10	**234**
10	**6688**
11	**0012224**
11	**5668899**
12	**00012344**
12	**5669**
13	**0034**
13	**79**

True-False: **1.** F **2.** F **3.** T **4.** F **5.** T **6.** F **7.** F **8.** T **9.** T **10.** F **11.** T **12.** F **13.** F.

Self-Quiz: **1.** e **2.** d **3.** d **4.** b **5.** d **6.** e **7.** a **8.** b **9.** c **10.** b **11.** a **12.** b **13.** a **14.** 21 **15.** 130 **16.** 26.62 **17.** 139 **18.** 98.56 **19.** 10 **20.** d **21.** 25.11% **22.** 52.88.

Chapter 4

Measures of Central Tendency and Variability

CHAPTER OUTLINE

I. **Measures of Central Tendency**. There are three commonly used measures of central tendency.

A. Arithmetic mean. Symbolized by "$\overline{X}$" (read X bar), or "μ" (read mu). This is simply the average of a set of raw scores. When referring to an entire population, the symbol used is μ. "$\overline{X}$" is used for sample data.

1. Equation for calculating the mean of sample raw scores

$$\overline{X} = (\sum X_i) / N$$

This equation simply tells us to add all of the scores and divide by the total number of scores.

a. The mean is sensitive to the exact value of all the scores in the distribution.
b. $\sum(X-\overline{X})=0$. The sum of the deviations about the mean equals zero.
c. The mean is very sensitive to extreme scores when the scores are not balanced at both ends of the distribution.
d. $\sum(X-\overline{X})^2$ is a minimum. The sum of the squared deviations of all the scores about their mean is a minimum.
e. Under most circumstances, of the measures used for central tendency, the mean is least subject to sampling variation.

2. Overall mean equation.

$$\overline{X}_{\text{overall}} = \frac{\text{sum of all scores}}{N} = \frac{n_1\overline{X}_1 + n_2\overline{X}_2 + \cdots + n_k\overline{X}_k}{n_1 + n_2 + \cdots + n_k}$$

which signifies that the overall mean is equal to the sum of the mean of each group times the number of scores in the group, divided by the sum of the number of scores in each group.

B. <u>The median (Mdn)</u> This is the value below which 50% of the scores fall. It is the same thing as P_{50}.

1. Equation for grouped scores:

$$\text{Mdn} = X_L + (i/f_i)(\text{cum } f_P - \text{cum } f_L)$$

2. Calculation with raw scores. The median is the centermost score, if the number of scores is odd. If the number is even, the median is taken as the average of the two centermost scores. (Note: scores are rank ordered first.)

3. Properties of the median:

   a. The median is less sensitive than the mean to unbalanced extreme scores.
   b. Under usual circumstances, the median is more subject to sampling variability than the mean, but less subject to sampling variability than the mode.

C. The mode. This is simply the most frequent score in the distribution. It is calculated by inspection.

D. Symmetry. In a unimodal (one mode) symmetrical distribution, the mean, median and mode are all equal. In a positively skewed distribution, the mean will be larger than the median. In a negatively skewed distribution, the mean will be lower than the median.

II. **Measures of Variability**. These measures quantify the extent of dispersion in a distribution.

A. Range. The range is the difference between the highest and lowest score in the distribution. The range measures only the spread of the two extreme scores.

B. Deviation scores. These tell how far away the raw score is from the mean of its distribution. It is symbolized by $(X - \overline{X})$ or $(X - \mu)$.

$$(X - \overline{X}) \quad \text{sample data}$$

$(X - \mu)$ population data

C. The standard deviation. This is a measure of an average deviation of raw scores about the mean. It is symbolized by "s" for the sample standard deviation, and by "σ" (small sigma) for the population standard deviation.

1. Conceptual formula:

$$s = \sqrt{\frac{\sum(X - \overline{X})^2}{N - 1}} = \sqrt{\frac{SS}{N - 1}}$$

2. Computational formula:

$$s = \sqrt{\frac{\sum X^2 - \frac{(\sum X)^2}{N}}{N - 1}}$$

because $\sum(X - \overline{X})^2 = \sum X^2 - \left(\sum X\right)^2 / N$

3. For the population standard deviation, the formula is just a little different. The denominator of the formula is N, instead of $N - 1$.

a. Conceptual formula:

$$\sigma = \sqrt{\frac{\sum(X-\mu)^2}{N}} = \sqrt{\frac{SS_{pop}}{N}}$$

b. Computational formula:

$$\sigma = \sqrt{\frac{\sum X^2 - \frac{(\sum X)^2}{N}}{N}}$$

4. Properties of the standard deviation.

   a. The standard deviation gives a measure of dispersion relative to the mean.
   b. The standard deviation is sensitive to each score in the distribution.
   c. It is stable with regard to sampling fluctuations.

D. The variance. This is simply the square of the standard deviation. The variance is symbolized as "s2" for the sample, and "σ2" for the population. This is read as "the variance" or "s squared."

1. Conceptual formula:

$$s^2 = \frac{\sum(X-\overline{X})^2}{N-1} = \frac{SS}{N-1} \quad \text{sample data}$$

$$\sigma^2 = \frac{\sum(X-\mu)^2}{N} = \frac{SS_{pop}}{N} \quad \text{population data}$$

2. Computational formula:

$$s^2 = \frac{\sum X^2 - \frac{(\sum X)^2}{N}}{N-1} \quad \text{sample data}$$

$$\sigma^2 = \frac{\sum X^2 - \frac{(\sum X)^2}{N}}{N} \quad \text{population data}$$

CONCEPT REVIEW

Organizing and presenting data is often a problem for scientists. We have just discussed frequency distributions. Frequency distributions do not allow (1) _______. statements about the distributions. Nor do they allow quantitative (2) _______ to be made between two distributions. Two characteristics of distributions often described are the (3) _______ tendency and (4) _______. Central tendency

(1) quantitative

(2) comparisons

(3) central

(4) variability

refers to how scores tend to (5) _______ around	(5) cluster
a central point in the distribution. Variability	
refers to the extent to which scores are	
(6) _______ or spread out.	(6) dispersed
There are three common measures of central	
tendency. They are the (7) _______,	(7) mean
(8) _______ and (9) _______. The	(8) median (9) mode
arithmetic mean is what we commonly compute	
when we compute the (10) _______ of a set of	(10) average
scores. The mean is the (11) _______ of the scores	(11) sum
divided by the (12) _______ of scores. We let	(12) number
the symbol (13) _______ stand for the mean of	(13) $\overline{X}$
sample scores. The equation for the mean of	
sample scores is $\overline{X}$ = (14) _______. For a	(14) $\Sigma X_i/N$
population mean we use the same equation	
except we let (15) _______ stand for the mean	(15) μ
in-stead of $\overline{X}$. The symbol μ is the Greek	
letter mu, pronounced "mew." Assume that	

you have observed the following scores

on a test: 23, 30, 17, 20, and 25. N

equals (16) _______.ΣX_i equals (17) _______, (16) 5
(17) 115

and the mean equals (18) _______. (18) 23

If you had the ages of all people at your

school and compiled the average age for the

school, you would be computing the population

mean,so you would use the symbol (19) _______. (19) μ

The mean possesses several interesting

properties.The first property of the mean is that

it is sensitive to the (20) _______ value of all of (20) exact

the scores in the distribution. This means if

one changes the value of any of the scores, this

will cause a (21) _______ in the value of the (21) change

mean,If one changes the value of more than

one score, the changes may cancel each other

out and the mean could stay the (22) _______. (22) same

But in general if you change (23) _______ the mean will (24) _______.

Another important property of the mean is that the sum of the deviations about the mean equals (25) _______. Written algebraically, this is (26) _______ = 0. This property says that if the (27) _______ is subtracted from each (28) _______, the sum of the (29) _______ will equal zero. $(X - \bar{X})$ is referred to as the (30) _______.

A third property of the mean is that it is very sensitive to (31) _______ scores when the scores are not (32) _______ at both ends of the distribution. For example, the mean of the numbers 8, 8, 9, 10, and 11 is (33) _______. The mean of 8, 8, 9, 10, 11 and 100 is (34) _______. We can see how adding one extreme score more than

(23) scores
(24) change
(25) zero
(26) $\Sigma (X - \bar{X})$
(27) mean
(28) score
(29) differences
(30) deviation score
(31) extreme
(32) balanced
(33) 9.2
(34) 24.33

doubles the mean.

A fourth property of the mean is that

the (35) _______ of the (36) _______ deviations (35) sum

of all the scores about their mean (36) squared
is a (37) _______.. The algebraic symbols (37) minimum

designating the sum of the squared deviations

of all scores about their mean is (38) _______. (38) $\Sigma (X - \bar{X})^2$

The final property of the mean we will

discuss is that under most circumstances,

of the measures of central tendency, the mean is

the (39) _______ subject to (40) _______ (39) least
(40) sampling

variation. This means that if we drew many

(41) _______ samples from a (42) _______, (41) random
(42) population

the means would probably be different for each

(43) _______ but the differences would be (43) sample

less for the mean than for the (44) _______ or (44) median

(45) _______. (45) mode

The overall mean for several groups

of scores is sometimes useful to calculate.

For k groups the overall mean is found by:

$$\overline{X}_{overall} = \frac{\text{sum of all scores}}{N} = \frac{n_1\overline{X}_1 + n_2\overline{X}_2 + \cdots + n_k\overline{X}_k}{n_1 + n_2 + \cdots + n_k}$$

This is true because $\overline{X}_1 = \Sigma X_1/n_1$ can be transformed

by multiplying the above equation by (46) _______. (46) n_1

This gives ΣX_1 (first group) = (47) _______. (47) $n_1\overline{X}_1$

This means that the overall mean is equal to the

sum of the (48) _______ of each group times (48) mean

the number of scores in the group, (49) _______ (49) divided

by the sum of the number of scores in each group.

If you are told that $\overline{X}_1 = 42$ and $n_1 = 20$, then ΣX_1

must equal (50) _______. Suppose you are told (50) 840

that the mean grade in one class was 78 for 250

students and 83 for 100 students in a second class.

The overall mean for the two classes equals:

$$\overline{X}_{overall} = \frac{(51)\_\_\_ \times 250 + (52)\_\_\_ \times 100}{(53)\_\_\_ + (54)\_\_\_} = (55)\_\_\_$$

(51) 78
(52) 83
(53) 250
(54) 100
(55) 79.43

The median is the value below which	
(56) _______% of the scores fall. It is the	(56) 50
same thing as the scale value at (57) _______.	(57) P_{50} or 50th percentile
For grouped data to calculate the (58) _______	(58) median
all one must do is calculate (59) _______. To	(59) P_{50}
calculate the median for raw scores you must	
first (60) _______ the scores. If the number of	(60) rank order
scores is odd, the median is the (61) _______	(61) center most
score. If the number of scores is even, the	
median is taken as the (62) _______ of the	(62) average
two centermost scores. So for the scores	
1, 2, 3, 4, 5, and 6, the median is (63) _______.	(63) (3 + 4)/2 = 3.5
Like the mean, the median has a couple	
of properties that are useful to know. First,	
the median is (64) _______ sensitive	(64) less
than the mean to unbalanced scores. For	

example, the mean of 1, 3 and 5 is (65) _______,	(65) 3
and the median of 1, 3 and 100 is (66) _______.	(66) 3
The second property of the median is that it is	
(67) _______ subject to sampling variability	(67) more
than the mean, but (68) _______ subject to	(68) less
sampling variability than the (69) _______.	(69) mode
The mode is the last measure of central	
tendency we will discuss. The mode is the	
(70) _______ score in the distribution.	(70) most frequent
Determining the mode is simply done	
by (71) _______. For the scores 90, 92, 93,	(71) inspection
96, 100, 100 and 100, the mode is (72) _______.	(72) 100
For the scores 1, 1, 1, 1, 2, 3, 9, 9, 9, and	
9, there are (73) _______ modes. The values	(73) two
for the modes are (74) _______ and	(74) 1
(75) _______.This is called	(75) 9
a (76) _______distribution. In a	(76) bimodal
frequency polygon,the mode is the (77) _______	(77) highest

point on the graph. If a distribution is

(78) _______ and (79) _______ the mean, — (78) unimodal

(79) symmetrical

median and mode are all equal. An example

of an unimodal, symmetrical distribution is the

(80) _______ curve. When the — (80) bell shaped

distribution is skewed, the median and mean

will not be equal. The (81) _______ is sensitive to — (81) mean

extreme scores. When the distribution

is negatively skewed the mean will — (82) less

be (82) _______than the median. When the

distribution is (83) _______ the — (83) positively skewed

mean will be greater than the median.

Measures of Variability

Variability has to do with how far scores

are (84) _______. Therefore, measures — (84) spread apart

of variability (85) _______ the extent — (85) quantify

of dispersion. There are three measures of

variability; the(86) _______, (87) _______,

(86) range
(87) standard Deviation

and the (88) _______. We have already

(88) variance

looked at the range. The range is the difference

between the (89) _______and

(89) highest

(90) _______scores in the distribution. So the

range of the scores 18, 20,

(90) lowest

27, 30, 32 and 41 is (91) _______ - (92) _______

(91) 41
(92) 18

= (93) _______. The range is easy to calculate

(93) 23

but of limited use because it measures

the spread of the two (94) _______

(94) extreme

scores, but not the spread of the scores

in between.

One might be more interested in how far

away scores are from the mean. A (95) _______

(95) deviation

score tells how far away the raw score ($\overline{X}$) is

from the mean (X). Mathematically this

is stated (96) _______.

(96) $X - \overline{X}$

The (97) _______ deviation uses the	(97) standard
(98) _______ scores of the data. The	(98) deviation
standard deviation is symbolized by an	
(99) _______ for sample data and the	(99) *s*
Greek letter (100) _______ for population data.	
The formula for the standard deviation	(100) σ
of a (101) _______ is	(101) population

$$\sigma = \sqrt{\frac{\sum (X - \mu)^2}{N}}$$

In words, this means subtract the value	
of the (102) _______ from each raw	(102) mean
score, square the result, sum each of the	
resulting (103) _______ scores, divide this sum	(103) squared deviation
by (104) _______, and then take the	(104) N
(105) _______.The reason we square the	(105) square root
$(X - \mu)$ term is because of one of the	
properties of the mean. Since $\Sigma\,(X - \mu)$	

equals (106) _______,	(106) zero
we can (107) _______ each $(X - \mu)$ term so	(107) square
the positive and negative differences don't	
cancel. To counteract squaring each term,	
we unsquare everything at the end by taking the	
(108) _______ of $\Sigma (X - \mu)^2/N$. Thus the	(108) square root
final equation for the	
standard deviation of the population is	

$$\sigma = \sqrt{\frac{\sum (X - \mu)^2}{N}}$$

The equation for the standard deviation	
of a (109) _______ uses (110) _______	(109) sample (110) $N - 1$
in the denominator rather than (111) _______	(111) N
to give a more accurate estimate of the	
population standard deviation. So the	
conceptual formula for the sample standard	
deviation is The numerator of the	

$$s = \sqrt{\frac{\sum (X - \overline{X})^2}{N - 1}} = \sqrt{\frac{SS}{N - 1}}$$

formula, $\sum (X - \overline{X})^2$, is also symbolized

by SS. SS stands for (112) _______. (112) sum of squares

Let's practice calculating "*s*" on the following sample data: 21, 24, 27, 30, and 33. Fill in the following table:

X	$X - \overline{X}$	$(X - \overline{X})^2$
21	**21 - 27**	**$(-6)^2$ = 36**
24	**24 - 27**	**$(-3)^2$ = 9**
27	**27 - 27**	**$(0)^2$ = 0**
30	**30 - 27**	**$(3)^2$ = 9**
33	**(113) ______**	**(114) ___ = ___ (115)**

(113) 33 - 27

(114) $(6)^2$

(115) 36

$\sum X$ = **(116)** _______; (116) 135

$\sum (X - \overline{X})^2$ = **(117)** _______; (117) 90

***N* = (118)** _______; (118) 5

$\overline{X} = \sum X / N$ = **(119)** _______ **= 27.0**

(119) 135/5

This (120) _______ formula is easy to use when the mean is (121) _______ (120) conceptual (121) not fractional. However, it can be shown that

$\Sigma (X - \overline{X})^2$ = (122) _______ which

(122) $\Sigma X^2 - (\Sigma X)^2/N$

allows a new formula for the computation

of s This computational formula is:

$$s = \sqrt{\frac{\sum X^2 - \frac{(\sum X)^2}{N}}{N-1}}$$

It is the preferred formula to use when

the mean is (123) _______.

(123) fractional

Applying the computational formula to

the same data, we obtain:

X	***X*²**
21	**441**
24	**576**
27	**(124) ___**
30	**900**
33	**1089**

(124) 729

ΣX = 135; ΣX^2 = (125) _____;
$(\Sigma X)^2$ = (126) _____;

(125) 3735
(126) 18,225

$$s = \sqrt{\frac{\sum X^2 - \frac{(\sum X)^2}{N}}{N-1}} = \sqrt{\frac{(127)\_\_\_ - \frac{(128)\_\_\_}{5}}{5-1}}:$$

(127) 3735
(128) 18,225

The standard deviation has certain characteristics. First, the standard deviation gives us a measure of dispersion relative to the (129) _______. Second, like the mean, the standard deviation is sensitive to (130) _______ in the distribution. Finally, like the mean, the standard deviation is (131) _______ with regard to sampling fluctuation.

(129) mean
(130) each score
(131) stable

The variance is the last measure of dispersion to be discussed. The variance of a set of scores is just the (132) _______ of the standard deviation.For a population we designate the variance as (133) _______. For a sample we again use (134) _______ as the denominator,

(132) square
(133) σ^2
(134) $N - 1$

because *N* gives too small an estimate for σ^2

when using sample data. Almost always,

one calculates the (135) _______ before the standard deviation since one has to take — (135) variance

the (136) _______ — (136) square root

of the variance to get the (137) _______. — (137) standard deviation

EXERCISES

1. The following IQ data has been obtained for 11 incoming graduate students:

   120, 130, 132, 139, 160, 115, 120, 142, 148, 120, 141

   Find each of the following statistics:

   a. mean
   b. median
   c. mode
   d. range

2. As a statistician (use your imagination), you are asked to recommend which measure of central tendency to use to characterize the following scores:

0.6, 0.2, 0.1, 0.2, 0.2, 0.4, 0.3, 0.7, 0.8, 0.1, 0.0, 0.5, 22.5, 0.4.

What is your recommendation and why?

3. An instructor decides to weight the scores on three tests differently. The first test is weighted only half as much as the second test. The third test is the final and it is weighted twice as much as the second test. The mean grade on the first test was 68, the mean of the second test was 75, and the mean on the third test was 80. What is the mean grade for all three tests, assuming the same number of students took all the exams?

4. What is the value of ΣX, if 900 students had a mean of 78.2 on a test?

5. Given the following sample scores on an aptitude test?

30, 20, 15, 19, 16

a. What is the mean of the scores?
b. What is the value of $\sum(X - \overline{X})$?
c. What is the standard deviation using the conceptual formula?
d. What is the value of $\sum(X - \overline{X})^2$?
e. What is the standard deviation using the computational formula?
f. What is the value of $\sum(X - \overline{X})^2$ using the computational formula?
g. What is the variance of these data?

6. Assume that the standard deviation of one distribution equals 15 and the standard deviation of another distribution is equal to 23. Which of the following distributions is more likely to

represent the case where $s = 15$ if the other distribution's standard deviation equals 23?

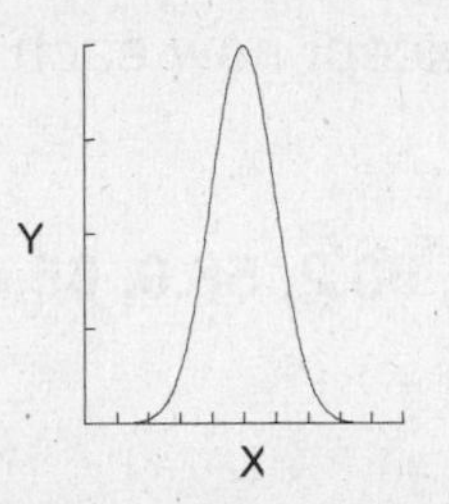

a. b.

7. Consider the following two sets of scores.

X: 80, 70, 55, 63, 72
Y: 50, 52, 59, 60

   a. Which of these sets of scores was more likely drawn from a population with the greater variance?
   b. Why?

8. Consider two more sets of scores. Notice that the Y scores are just the X scores with a constant value of 10 added to each score. What effect does adding a constant have on the value of the standard deviation? Do you see why?

X: 1, 3, 5, 7, 9
Y: 11, 13, 15, 17, 19

9. A car company wanted to know how many miles per gallon an experimental car got after a series of tests. The following data were obtained.

Miles per gallon: 31.2, 28.6, 36.4, 37.3, 30.1, 29.0, 32.7, 31.9

   a. What is the mean number of miles per gallon obtained in this sample?
   b. What is the value of ΣX and ΣX^2?

c. What is the standard deviation of this sample?
d. What is the variance of the sample?

10. Consider the same values as above except now each value is multiplied by 2. The results are:

Miles per gallon: 62.4, 57.2, 72.8, 74.6, 60.2, 58.0, 65.4, 63.8

a. What is the mean of this data?
b. What is the standard deviation?
c. What is the variance of the sample?

11. Repeat the above analysis on the original data in problem 9; this time multiply each original value by 4. The resulting data are:

Miles per gallon: 124.8, 114.4, 145.6, 149.2, 120.4, 116.0, 130.8, 127.6

a. What is the mean now?
b. What is the standard deviation?
c. What is the variance?
d. What is the relationship between the constant used to multiply the original data and the resulting mean, standard deviation and variance?

12. A statistics class taught by Dr. X had 30 students who obtained an average test score of 82.5. A class taught by Dr. Why had 20 students who ended up with an average score of 79.6. Dr. Zee (you guessed it) had 23 students who obtained an average score of 96.0. What was the average score for all the students combined?

TRUE-FALSE QUESTIONS

T F 1. Central tendency and variability are two terms that mean the same thing.

T F 2. The mean is the same as the arithmetic average.

T F 3. By changing the value of one score in a sample this will necessarily cause the mean to change.

T F 4. By changing the value of one score in a sample this will necessarily cause the median to change.

T F 5. By changing the value of one score in a sample this will necessarily cause the mode to change.

T F 6. The mean is the most sensitive measure of central tendency to extreme scores.

T F 7. $\sum(X-\overline{X}) = 0$

T F 8. $\sum(X-\overline{X})^2 = 0$

T F 9. Of the measures of central tendency, the mean is the least subject to sampling variations.

T F 10. If a group of 100 people has an average IQ of 107 and a group of 50 people has an average IQ of 101, the average IQ of all the people is 105.

T F 11. It is possible to have more than one mode in a sample of scores.

T F 12. The formula for the population standard deviation is

$$\sqrt{\sum (X - \overline{X})^2 / N - 1}$$

T F 13. Measures of variability quantify the extent of dispersion in a set of scores.

T F 14. The variance is the square of the standard deviation.

T F 15. The symbol *SS* stands for sum of squares.

T F 16. If one knows ΣX, N, and ΣX^2, one can calculate both the variance and the standard deviation. It is not necessary to know the mean itself.

T F 17. If one uses N instead of $N - 1$ in the denominator of the equation for the sample standard deviation, one will underestimate the population standard deviation.

T F 18. The equation used for sample standard deviation is an estimate of the population standard deviation.

SELF-QUIZ

1. The mean is _______ sensitive to extreme scores than the median.

   a. more
   b. less
   c. equally
   d. can't say without the scores

2. In a negatively skewed distribution, the mean will generally be _______ the median.

   a. greater than
   b. less than
   c. equal to
   d. can't tell from this information

3. $\sum(X - \overline{X})^2$ is called _______.

   a. standard deviation
   b. mean
   c. variance
   d. sum of squares
   e. confusing

4. If 5 points are added to each score in a distribution, which of the following will happen?

   a. the mean will be unchanged
   b. the standard deviation will be unchanged
   c. the mode will be unchanged
   d. the mean, median and mode will be unchanged

5. $\sum(X - \overline{X})$ equals _______ .

   a. the sum of squares
   b. 0
   c. 1
   d. cannot be determined

6. $\sum(X-\overline{X})^2$ equals _______ .

   a. 0
   b. 1
   c. the variance
   d. cannot be determined from the information provided

7. Which of the following is not a measure of central tendency?

   a. mode
   b. median
   c. mean
   d. standard deviation

8. $\sum(X-\overline{X})^2$ equals _______ .

   a. ΣX^2
   b. $\Sigma X^2 - (\Sigma X)^2/N$
   c. a and b
   d. cannot be determined from information provided
   e. sum of squares
   f. b and e

9. If $s = 12$ for a set of 200 scores and 5 points were subtracted from each of the scores, the new value of the standard deviation would be _______.

   a. 12/5
   b. $\sqrt{12/5}$
   c. 12 - 5
   d. 12

10. Assume one has a set of 35 scores with $\overline{X} = 100$ and $s = 9$. If 5 scores, each equal to 100 were added to the distribution, what effect would this have on $\overline{X}$?

    a. increase $\overline{X}$
    b. decrease $\overline{X}$
    c. $\overline{X}$ would remain unchanged
    d. cannot tell

11. Considering again the information in problem 10, what would be the effect of adding those same new scores on the value of s?

    a. increase s
    b. decrease s
    c. s would remain unchanged
    d. cannot be determined from the information given

12. What is the appropriate symbol for the formula, $\sqrt{\frac{\sum(X-\mu)^2}{N}}$

    a. s
    b. s^2
    c. σ
    d. σ^2

13. Consider a distribution with $N = 100$, $\overline{X} = 50$, and $\sum(X-\overline{X})^2 = 9875$. If 10 new scores of 50 were added to the distribution, what wou1d happen to the sum of squares?

    a. the value would increase
    b. the value would decrease
    c. the value would remain unchanged
    d. cannot be determined from information given

For questions 14 - 22 use the following sample data:

X: 10, 12, 6, 8, 9, 11, 13, 13, 5, 0, 1

14. What is the mean?

    a. 8.0
    b. 11.0
    c. 13.5
    d. 6.0

15. What is the median?

    a. 8
    b. 9
    c. 10
    d. 11

16. What is the mode?

    a. 0
    b. 6
    c. 13
    d. 9

17. What is the range?

    a. 0
    b. 13
    c. 1
    d. 3.6

18. What is the variance?

    a. 20.6
    b. 4.54
    c. 18.7

d. 4.3

19. What is the value of $\sum(X-\overline{X})$?

    a. 4.54
    b. 206
    c. 1
    d. 0

20. What is the value of $\sum(X-\overline{X})^2$?

    a. 4.54
    b. 206
    c. 1
    d. 0

21. If the above data were a population set of scores, what would the standard deviation be?

    a. 4.33
    b. 6.92
    c. 4.54
    d. 10

22. The mean number of students in a classroom at school A is 32.5 and there are 25 classrooms. The mean number of students at school B is 29.6 in 12 classrooms. At school C the mean in each class is 15.3 for 10 classrooms. What is the mean number of students per classroom for all the schools combined?

    a. 1320.7
    b. 28.1
    c. 25.7
    d. 28.7

ANSWERS

Exercises: **1a.** 133.36 **1b.** 132 **1c.** 120 **1d**. 45 **2.** median. Mean would be too influenced by value 22.5. Since there's no impressive mode in the data, the median would appear to be the best choice. **3.** 76.86 **4.** 70,380 **5a.** 20.0 **5b.** 0 **5c.** 5.96 **5d.** 142 **5e.** 5.96 **5f.** 142 **5g.** 35.5 **6.** a **7a.** *X* **7b.** The standard deviation of the *X* scores is 9.46 while the standard deviation of the *Y* scores is only 4.99 suggesting that *Y* may have been drawn from a population with less variance than *X*. **8.** no effect on s because *X* increases by the same constant that each individual score does so the value of $\Sigma\ (X - X)^2$ does not change **9a.** 32.15 **9b.** $\Sigma\ X = 257.2$; $\Sigma\ X^2 = 8{,}341.56$ **9c.** 3.22 **9d**. 10.37 **10a.** 64.30 **10b.** 6.44 **10c.** 41.47 **11a.** 128.6 **11b.** 12.88 **11c.** 165.90 **11d.** The new mean is equal to the original mean times the constant. The same is true of the new standard deviation. The new variance is equal to the original variance times the square of the constant (within our rounding errors). **12.** 85.96.

True-False: **1.** F **2.** T **3.** T **4.** F **5.** F **6.** T **7.** T **8.** F **9.** T **10.** T **11.** T **12.** F **13.** T **14.** T **15.** T **16.** T **17.** T **18.** T

Self-Quiz: **1.** a **2.** b **3.** d **4.** b **5.** b **6.** d **7.** d **8.** f **9.** d **10.** c **11.** b **12.** c **13.** c **14.** a **15.** b **16.** c **17.** b **18.** a **19.** d **20.** b **21.** a **22.** b.

Chapter 5

THE NORMAL CURVE AND STANDARD SCORES

CHAPTER OUTLINE

I. The Normal Curve

A. Important in behavioral sciences.

1. Many variables of interest are approximately normally distributed.
2. Statistical inference tests have sampling distributions which become normally distributed as sample size increases.
3. Many statistical inference tests require sampling distributions that are normally distributed.

B. Characteristics.

1. Symmetrical, bell-shaped curve.
2. Equation.

$$Y = \frac{N}{\sigma\sqrt{2\pi}} e^{-(X-\mu)^2/2\sigma^2}$$

Shows that the curve is asymptotic to the abscissa, i.e., it approaches the *X* axis and gets closer and closer but never touches it.

3. Area contained under the normal curve:

   a. Area under the curve represents the percentage of scores contained within the area.
   b. 34.13% of scores between mean (μ) and $+1\sigma$, 13.59% of area contained between a score equal to $\mu + 1\sigma$ and a score of $\mu + 2\sigma$, 2.15% of area is between $\mu + 2\sigma$ and $\mu + 3\sigma$, and 0.13% falls beyond $\mu + 3\sigma$.
   c. Since the curve is symmetrical, the same percentages hold for scores below the mean.

II. Standard Scores (*z* Scores)

A. Symbol. Symbolized as *z*.

B. Definition. A standard score is a transformed score which designates how many standard deviation units the corresponding raw score is above or below the mean.

C Equation.

$$z = (\overline{X} - \mu)/\sigma \quad \text{population data}$$

$$z = (X - \overline{X})/s \quad \text{sample data}$$

D. Comparisons between different distributions.

1. Allows comparisons even when the units of the distributions are different.
2. Percentile ranks are possible.

E. Characteristics of *z* scores.

1. *z* scores have the same shape as the set of raw scores from which they were transformed.
2. $\mu_Z = 0$. The mean of *z* scores equals zero.
3. $\sigma_Z = 1.00$. The standard deviation of *z* scores equals 1.00.

F. Using *z* scores.

1. Finding the area given the raw score.

$$z = (X - \mu)/\sigma$$

Use above formula to calculate z score. Then use table to determine the area under the normal curve for the various values of z.

2. Finding the raw score given the area.

$$X = \mu + \sigma z$$

Use above formula substituting the value of *z* that designates the area under the curve one wishes, and solve for *X*, the raw score.

CONCEPT REVIEW

The (1) _______ curve is extremely	(1) normal
important in the behavioral sciences. Many	
(2) _______ measured have distributions	(2) variables
which closely approximate the normal curve	
. Many (3) _______ tests have	(3) inference
(4) _______ distributions which become	(4) sampling
(5) _______ distributed as(6) _______	(5) normally (6) sample
size (7) _______. Also, many inference	(7) increases
tests require sampling distributions which	
are normally distributed.	
The normal curve has certain	
characteristics. It is a (8) _______	(8) symmetrical
, (9) _______ - (10) _______ curve.	(9) bell (10) shaped
As the curve approaches the (11) _______ axis,	(11) *X* or horizontal
it is changing the *Y* value very (12) _______	(12) slowly
The curve (13) _______ the *X* axis and	(13) approaches

gets closer to it, but never touches it. The curve is said to be (14) _______ to the X axis	(14) asymptotic.
In a normal curve, there is a special relationship	
between the (15) _______ and (16) _______	(15) mean standard deviation
with regard to the (17) _______ contained under	(17) area
the curve. When a set of scores is (18) _______	(18) normally
distributed, (19) _______ % of the	(19) 34.13
area under the curve is contained between the	
(20) _______ and a score which is equal to	(20) mean
(21) _______ + (22) _______; (23) _______ %	(21) μ (22) 1σ (23) 13.59
of the area is contained between a score equal to	
(24) _______ +(25) _______	(24) μ (25) 1σ
and a score of (26) _______ + (27) _______;	(26) μ (27) 2σ
(28) _______ % of the area is contained	(28) 2.15
between (29) _______ + (30) _______	(29) μ (30) 2σ
and (31) _______ + (32) _______;	(31) μ

	(32) 3σ
and (33) _______ %	(33) 0.13
of the area exists beyond (34) _______ +	(34) μ
(35) _______.	(35) 3σ
Since the curve is symmetrical, the same	
percentages hold or scores (36) _______ the mean.	(36) below
In graphing a normal curve (37) _______ is	(37) frequency
plotted on the vertical axis, and the percentages	
above represent the (38) _______ of (39) _______	(38) percentage (39) scores
contained within the area.	
One can use this information to determine	
the frequency of scores above or below a given	
score. For example, (40) _______% of the	(40) 50
scores would fall below the mean of a	
normal distribution.	

Standard Scores

The main reason for standard scores is to allow one to	
(41) _______ a score against a reference group.	(41) compare
A *z* standard score is a (42) _______ score	(42) transformed

which designates how many (43) _______ — (43) standard deviation

units the corresponding (44) _______ — (44) raw

score is above or below the (45) _______. — (45) mean

In equation form, $z =$ ((46) _______ — (46) X

– (47) _______)/(48) _______ — (47) μ (48) σ

for population data. In conjunction with the

normal curve the z score allows one to determine

the (49) _______ or (50) _______ of scores — (49) number (50) percentage

which fall above or below any score in the

distribution. Consider a population set of scores

where $\mu = 30$ and $\sigma = 5$.

A raw score value of 40 would have a value

of $z =$ ((51) _______ – (52) _______)/ — (51) 40 (52) 30

(53) _______ =(54) _______. Therefore, a raw — (53) 5 (54) 2.00

score value of 40 would have a percentile rank of

(55) _______. An important use of z — (55) 97.72

scores is to be able to (56) _______ — (56) compare

scores that are not otherwise directly comparable.

By converting your height and weight to

(57) _______ scores, you could compare — (57) z

your (58) _______ standing on height and — (58) relative

weight.

There are three characteristics of z scores

worth remembering. First, z scores have the

same (59) _______ as the set — (59) shape

of raw scores. Second, the mean of the z scores

(symbolized by (60) _______) equals (61) _______ — (60) μ_z (61) 0

Third, the (62) _______ — (62) standard deviation

of z scores equals (63) _______. — (63) 1.00

There is a (64) _______ of z scores which — (64) table

gives the (65) _______ of area — (65) proportion

under the standard normal curve for any (Table A)

given (66) _______ score. In the text — (66) z

the first column (A) of the z table contains

the (67) _______ score. Column B lists	(67) *z*
the (68) _______ of the total area between	(68) proportion
a given *z* score and the (69) _______.	(69) mean
Column C lists the proportion of the total	
area which exists (70) _______ the *z* score.	(70) beyond
One can use the table to	
find the (71) _______ rank of a given	(71) percentile
(72) _______ score. Assume that a	(72) raw
college entrance exam was normally distributed	
with $\mu = 400$ and $\sigma = 100$. To deter-	
mine the percentile rank of a score of 450 you must	
first calculate the (73) _______ (74) _______.	(73) *z* (74) score
To do this subtract (75) _______	(75) 400
from (76) _______ and divide the result by	(76) 450
(77) _______. The result is $z =$ (78) _______	(77) 100 (78) 0.50
The next step is to find the (79) _______	(79) proportion
of scores less than of a *z* score of (80) _______	(80) 0.50
Going to Column B of the table, we find the proportion	

of area between the (81) _______ and a value	(81) mean
of 0.50 equals (82) _______	(82) 0.1915
To this value we must add (83) _______	(83) .5000
to take into account the area (84) _______	(84) below
the mean. Thus, the proportion of scores	
below a college entrance exam	
score of 450 is (85) _______ + (86) _______	(85) 0.5000 (86) 0.1915
= 0.6915. To convert this proportion to a	
percentile rank, (87) _______ the (88) _______.	(87) multiply (88) 100
The percentile rank of 450 is (89) _______.	
Another way to solve the same	
problem is to go to Column (90) _______	(89) 69.15 (90) C
which is the area (91) _______ the z score and	(91) beyond
subtract that value from (92) _______	(92) 1.00
and then multiply by (93) _______. This	(93) 100
method works only when the raw score is	
greater than the (94) _______.	(94) mean

If one had a z score below the mean it

would have a (95) _______ value. The table (95) negative

only gives (96) _______ values. (96) positive

Since the normal curve is symmetrical, to find

the percentile rank of a negative z score, one

finds the (97) _______ beyond (97) area

the corresponding (98) _______ z score and (98) positive

multiplies by (99) _______. (99) 100

Sometimes one wishes to find the raw score

at a given percentile from a normal distribution.

Considering again the college entrance

test, ($\mu = 400$; $\sigma = 100$), how does one

determine the raw score at the 75th percentile?

In the z equation the variable

(100) _______ stands for the raw score. (100) X

We must do some algebra and solve for X.

The equation for the

raw score of a percentile

rank is X = (101) _______ + | (101) μ

(102) _______ x (103) _______. An easy | (102) σ
(103) z

way to solve such a problem is to recognize that

at the 75th percentile the proportion of the

area beyond this point is

(104) _______. Going to | (104) 0.2500

Column C of the table, we locate the area

closest to 0.2500 and note this point has a *z* score

equal to (105) _______ in | (105) 0.67

Column A. Solving the equation we have

X = (106) _______ + (107) _______ x | (106) 400
(107) 100

(108) _______, which equals (109) _______. | (108) 0.67
(109) 467

EXERCISES

For problems 1 through 8 use the following information:

In a population survey of patients in a rehabilitation hospital, the mean length of stay in the hospital was 12.0 weeks with a standard

deviation equal to 1.0 week. The distribution was normally distributed.

1. Out of 100 patients how many would you expect to stay longer than 13 weeks?

2. What is the percentile rank of a stay of 11.3 weeks?

3. What percentage of patients would you expect to stay between 11.5 weeks and 13.0 weeks?

4. What percentage of patients would you expect to be in longer than 12.0 weeks?

5. How many times out of 10,000 would you expect a patient selected at random to remain in the hospital longer than 14.6 weeks?

6. What proportion of patients are likely to be in less than 9.7 weeks?

7. What is the length of stay at the 90th percentile?

8. What is the length of stay at the 50th percentile?
9. On one college aptitude test with a mean of $\mu = 100$ and a standard deviation of $\sigma = 16$, a student achieved a score of 124. The same student took a different test which had a mean of $\mu = 50$ and a standard deviation of $\sigma = 10$. On the second test the student achieved a score of 65. On which test did the student do better?

10. If the mean height of college males is 70 inches with a standard deviation of 3 inches, what percentage of college males would be between 6' and 6'4"? Assume a normal distribution.

11. Using the information in problem 10, what height would someone have to be in order to be in the 99th percentile?

12. Using the data in problem 10, what is the height below which the shortest 2.5% of the college males fall and what is the height above which the tallest 2.5% fall?

13. A surgeon is experimenting with a new technique for implanting artificial blood vessels. Using this technique with a great many operations, the mean time before clotting of an artificial blood vessel has been 32.5 days with a standard deviation of 2.6 days. The following data were obtained on four operations.

Operation	Time to Clot
A	40
B	31
C	29
D	36

a. What are the *z* scores for the four operations?
b. What is the percentile rank for each of the four operations? Assume a normal distribution.
c. How long would a vessel have to stay open to be in the 95th percentile? Assume a normal distribution.

14. Given the following *z* scores, find the area below *z*:

a. 1.68
b. –0.45
c. –1.96
d. –0.52
e. 2.58

15. Assuming that you wished to have the highest possible score on an exam relative to the other scores; would you rather have a score of 70 on a test with a mean of 60 and a standard deviation of 5.2 or a score of 81 on a test with a mean of 70 and a standard deviation of 7.1?

16. What is the percentile rank for each of the following *z* scores? Assume a normal distribution.

    a. 1.23
    b. 0.89
    c. –0.46
    d. –1.00

17. What *z* scores correspond to the following percentile ranks? Assume the scores are normally distributed.

    a. 50
    b. 46
    c. 96
    d. 75
    e. 34
    f. 4

TRUE-FALSE QUESTIONS

T F 1. The normal curve is a symmetric, bell-shaped curve.

T F 2. The area under the normal curve represents the percentage of scores contained within the area.

T F 3. It is impossible to have a *z* score of 23.5.

T F 4. A z transformation will allow comparisons to be made when units of distributions are different.

T F 5. If the original raw score distribution is not normally distributed, the mean of the z transformation scores of the raw data will not equal 0.

T F 6. In the standard normal curve, 13.59% of the scores will always be contained between the mean (μ) and $+1\sigma$.

T F 7. In a plot of the normal curve, frequency is plotted on the X axis.

T F 8. The standard deviation of the z distribution is always equal to 1.0.

T F 9. The area beyond a z score of +2.58 is 0.005.

T F 10. One cannot reasonably do z transformations on ratio data.

T F 11. The normal curve never touches the X axis.

T F 12. To do a z transformation, one must know only the population mean and the value of the raw score to be transformed.

T F 13. To calculate the score at the 97.5th percentile, one would apply the formula $X = \mu + (\sigma)(1.96)$.

T F 14. The z score and the z distribution are the same thing.

SELF-QUIZ

1. If a distribution of raw scores is negatively skewed, transforming the raw scores into z scores will result in a _______ distribution.

   a. normal
   b. bell-shaped
   c. positively skewed
   d. negatively skewed

2. The mean of the z distribution equals _______.

   a. 0
   b. 1
   c. N
   d. depends on the raw scores

3. The standard deviation of the z distribution equals _______.

   a. 0
   b. 1
   c. the variance of the z distribution
   d. b and c

4. $\Sigma(z - \mu_z)$ equals _______.

   a. 0
   b. 1
   c. the variance
   d. cannot be determined

5. The proportion of scores less than $z = 0.00$ is _______.

   a. 0.00
   b. 0.50
   c. 1.00
   d. –0.50

6. In a normal distribution the *z* score for the mean equals _______.

   a. 0
   b. the *z* score for the median
   c. the *z* score for the mode
   d. all of the above

7. In a normal distribution approximately _______ of the scores will fall within 1 standard deviation of the mean.

   a. 14%
   b. 95%
   c. 70%
   d. 83%

8. Would you rather have an income (assume a normal distribution and you are greedy) _______.

   a. with a *z* score of 1.96
   b. in the 95th percentile
   c. with a *z* score of -2.00
   d. with a *z* score of 0.000

9. How much would your income be if its *z* score value was 2.58?
   a. $10,000
   b. $ 9,999
   c. $ 5,000
   d. cannot be determined from information given

10. Which of the following *z* scores represent(s) the most extreme value in a distribution of scores assuming they are normally distributed?

    a. 1.96
    b. 0.0001
    c. –0.0002
    d. –3.12

11. What is the percentile rank of a *z* score of –0.47?

    a. 31.92
    b. 18.08
    c. 50.00
    d. 47.00
    e. 0.06

12. A standardized test has a mean of 88 and a standard deviation of 12. What is the score at the 90th percentile? Assume a normal distribution.

    a. 90.00
    b. 112.00
    c. 103.36
    d. 91.00

13. On a test with a population mean of 75 and standard deviation equal to 16, if the scores are normally distributed, what is the percentile rank of a score of 56?

    a. 58.30
    b. 0.00
    c. 25.27
    d. 38.30
    e. 11.70

14. On the test referred to in problem 13, what percentage of scores fall below a score of 83.8?

    a. 55.00
    b. 79.12
    c. 20.88
    d. 29.12
    e. 70.88

15. Using the same data as in problem 13, what percentage of scores fall between 70 and 80?

    a. 75.66
    b. 70 23
    c. 24.34
    d. 23.57
    e. 12.17

ANSWERS

Exercises: **1.** 16 to nearest whole number **2.** 24.20% **3.** 53.28% **4.** 50% **5.** 47 **6.** 0.0107 **7.** 13.28 weeks **8.** 12 weeks **9.** the student did equally well on both tests **10.** 22.86% **11.** 76.99 inches **12.** shortest, less than 64.12 inches; tallest, greater than 75.88 inches **13a.** A = 2.88; B = -0.58; C = -1.35; D = 1.35 **13b.** A = 99.80%; B = 28.10%; C = 8.85%; D = 91.15% **13c.** 36.78 days (using $z = 1.645$) **14a.** 0.9535 **14b.** 0.3264 **14c.** 0.0250 **14d.** 0.3015 **14e.** 0.9951 **15.** 81 has a z score of 1.55, while the 68 has a z score of 1.54. The 70 is the higher score relative to the other. **16a.** 89.07 **16b.** 81.33 **16c.** 32.28 **16d.** 15.87 **17a.** 0.00 **17b.** –0.10 **17c.** 1.75 **17d.** 0.67 **17e.** –0.41 **17f.** –1.75.

True-False: **1.** T **2.** T **3.** F **4.** T **5.** F **6.** F **7.** F **8.** T **9.** T **10.** F **11.** T **12.** F **13.** T **14.** F.

Self-Quiz: **1.** d **2.** a **3.** d **4.** a **5.** b **6.** d **7.** c **8.** a **9.** d **10.** d **11.** a **12.** c **13.** e **14.** e **15.** c.

Chapter 6

Correlation

CHAPTER OUTLINE

I. Relationships

A. Linear Relationships.

1. Scatter plots. A scatter plot is a graph of paired *X* (one variable score) and *Y* (another variable score) values. By visually examining the graph one can get a good idea of the nature of the relationship between the two variables (i.e., linear or not).

2. Definitions.

   a. Linear relationship. A linear relationship between two variables is one in which the relationship between two variables can accurately be represented by a straight line.
   b.. Curvilinear relationship. When a curved line fits a set of points better than a straight line it is called a curvilinear association or relationship.

3. Straight Line Equation.

a. General equation.
$$Y = bX + a$$

where a = the Y intercept and b = the slope of the line.

b. Slope of the straight line equation (b). The slope tells us how much the Y score changes for each unit change in the X score. In equation form:

$$b = \textbf{slope} = (Y_2 - Y_1)/(X_2 - X_1)$$

The slope is a constant value.

c. Y intercept (a). The Y intercept is the value of Y where the line intersects the Y axis. It is the value of Y when $X = 0$.

B. Positive relationships. This indicates that there is a direct relationship between the variables. Higher values of X are associated with higher values of Y and vice versa.

C. Negative relationships. This exists when there is an inverse relationship between X and Y. Low values of X are associated with high values of Y and vice versa.

D. Perfect relationship. This occurs when all the pairs of points fall on a straight line.

E. Imperfect relationships. This is when a positive or negative, relationship exists but all of the points do not fall on the line.

II. Correlation Concepts

A. <u>Definition</u>. Correlation is a measure of the direction and degree of relationship that exists between two variables.

B. <u>Correlation coefficient</u>. Expresses quantitatively the magnitude and direction of the correlation.

1. Range. Can range from +1 to –1.
2. Sign. The sign of the coefficient tells us whether the relationship is positive or negative.
3. Magnitude. The coefficient ranges from +1 to –1. Plus 1 is a perfect positive correlation, and minus 1 expresses a perfect negative relationship. A zero value of the correlation coefficient means there is no relationship between the two variables. Imperfect relationships vary between 0 and 1. They will be plus or minus depending on the direction of the relationship.

III. Pearson *r*

A. <u>Definition</u>.

Pearson *r* is a measure of the extent to which paired scores occupy the same position within their own distributions. Standard scores allow us to examine the relative positions of variables independent of the units of measure.

B. <u>Calculating *r*</u>.

1. Computational formula from raw scores:

$$r = \frac{\sum XY - \frac{(\sum X)(\sum Y)}{N}}{\sqrt{\left[\sum X^2 - \frac{(\sum X)^2}{N}\right]\left[\sum Y^2 - \frac{(\sum Y)^2}{N}\right]}}$$

III. Additional Interpretation for Pearson r

A. Variability of Y. Pearson r can also be interpreted in terms of the variability of Y accounted for by X.

1. $r = 0$. Where $r = 0$, knowledge of X does not help us predict Y. Best prediction of Y when $r = 0$ is $\overline{Y}$.

2. Deviation of Y_i. Distance between a given score Y_i and the mean of Y scores $(\overline{Y})$ is divisible into two parts.

Deviation of Y_i = Error in prediction + deviation of Y_i accounted for by X

$$(Y_i - \overline{Y}) = (Y_i - Y') + (Y' - \overline{Y})$$

3. Total variability.

$$\Sigma (Y_i - \overline{Y})^2 = \Sigma(Y_i - Y')^2 + \Sigma(Y' - \overline{Y})^2$$

As the relationship between X and Y gets stronger, the prediction error gets smaller causing $\Sigma(Y_i - Y')^2$ to decrease, and $\Sigma(Y' - \overline{Y})^2$ to increase.

B. New definition of r. Pearson r equals the square root of the proportion of the total variability of Y accounted for by X.

C. Explained Variability. The explained variability = r^2. For example, if $r = .7$ then .49 or 49% of the variability of Y is accounted for by X. This is called the explained variability. If X is causal with respect to Y, r^2 is also a measure of the size of the effect.

IV. Other correlation coefficients besides *r*

A. Eta (η). This is used for curvilinear relationships where Pearson r would underestimate the degree of relationship.

B. Biserial correlation coefficient (r_b). Used when one variable is measured on an interval scale and the other variable is dichotomous.

C. Phi coefficient (Φ). Used when both variables are dichotomous.

D. Spearman rank order correlation coefficient (r_S), also called rho. Used when one or both variables are of ordinal scaling.

1. Computational equation.

$$r_s = 1 - \frac{6\sum D_i^2}{N^3 - N}$$

where D_i = difference between ith pair of ranks
N = number of pairs of ranks

2. Uses. When the data are not of either interval or ratio scaling but are of ordinal scaling, r_s can be used.

V. Correlation and Causation

A. Causation. Correlation between X and Y does not prove causation.

B. Explanations for correlation between X and Y.

1. Correlation may be spurious
2. X causes Y
3. Y causes X
4. Third variable causes the correlation between X and Y

C. Role of experimentation. To establish that one variable is the cause of another, an experiment must be conducted by systematically varying only the causal variable and then measuring the effect on the other variable.

CONCEPT REVIEW

The easiest way to determine if a relationship	
exists between two variables is to (1) _______	(1) plot
the variables on a (2) _______.	(2) graph
Such a plot is called a (3) _______ (4) _______	(3) scatter (4) plot
A scatter plot is a graph of (5) _______	(5) paired
X and *Y* scores. When a (6) _______	(6) straight
line accurately describes the relationship	
between two variables,the relationship is called	
(7) _______.Not all relationships are linear.	(7) linear
Those that are not are called	
(8) _______. In these cases a	(8) curvilinear
(9) _______ line fits the	(9) curved
(10) _______ better than a	(10) points
(11) _______ line. Although	(11) straight
a (12) _______ solution is sometimes used for	(12) graphic
prediction, it is more common to predict	

(13) _______ from the (14) _______	(13) Y (14) equation
of the straight line. The general form of the	
equation for a straight line is: Y = (15) _______	(15) b
times (16) _______ + (17) _______,	(16) X (17) a
where (18) _______ = the Y intercept and	(18) a
(19) _______ = the slope of the line.	(19) b
The (20) _______ is the value of Y where	
the line intersects the	(20) Y intercept
(21) _______.	(21) Y axis or ordinate
(22) _______	(22) Y
when X = (23) _______	(23) 0
The slope of a line measures its	
(24) _______ of (25) _______.	(24) rate (25) change
The slope tells how much the (26) _______	(26) Y
score changes for each (27) _______	(27) unit
change in the (28) _______	(28) X

score. In straight line functions, the slope has a

(29) _______ (29) constant

value for any points on the line.

In conceptual terms the equation for the

slope is:

$$b = \frac{\Delta(30)\_\_\_\_}{\Delta(31)\_\_\_\_} = \frac{(32)\_\_\_\_ - (33)\_\_\_\_}{(34)\_\_\_\_ - (35)\_\_\_\_}$$

(30) Y
(31) X
(32) Y_2
(33) Y_1
(34) X_2
(35) X_1

If one had two pairs of points (10, 20) and

(15, 30), the slope for the line

connecting these points would be:

$$b = \frac{(36)\_\_\_\_ - (37)\_\_\_\_}{(38)\_\_\_\_ - (39)\_\_\_\_} = (40)\_\_\_\_$$

(36) 30
(37) 20
(38) 15
(39) 10
(40) 2

Relationships between two variables may

be either (41) _______ or (42) _______

(41) positive
(42) negative

If the relationship is positive, the slope is	
(43) _______. If the slope is (44) _______	(43) positive (44) negative
the relationship is (45) _______.	(45) negative
In a positive relationship higher values of	
(46) _______ are associated with	(46) X
(47) _______ values of	(47) higher
(48) _______. In a negative	(48) Y
relationship (49) _______ values of	(49) lower
(50) _______ are associated with	(50) X
(51) _______ values of (52) _______. On a	(51) higher (52) Y
graph, a negative slope would run down-ward	
from (53) _______ to (54) _______.	(53) left (54) right
In a negative relationship as X (55) _______, Y	(55) increases
(56) _______. In a positive relation	(56) decreases
ship, as X (57) _______ Y (58) _______.	(57) increases (58) increases
(59) _______ and	(59) Correlation
(60) _______ are closely related. Correlation	(60) regression

is used to determine whether a (61) _______

(61) relation-ship

exists between two variables and to establish

its (62) _______ and (63) _______.

(62) magnitude
(63) direction

The degree of relationship may vary from being

(64) _______ to (65) _______.

(64) non-existent
(65) perfect

When the relationship is perfect, the

correlation is at its (66) _______ and

(66) highest

we can exactly (67) _______

(67) predict

one variable from the other. As *X* changes, so

does (68) _______. The points will all fall on a

(68) *Y*

(69) _______ (70) _______. When a relationship

(69) straight
(70) line

is nonexistent, the correlation is (71) _______,

(71) lowest

and knowing the value of one variable doesn't help

at all in (72) _______ the other. (73) _______

(72) predicting
(73) Imperfect

relationships have intermediate levels of

correlation and (74) _______ is approximate.	(74) prediction
The same value of X doesn't always lead	
to the (75) _______value of Y. But on the	(75) same
average, Y changes (76) _______ with	(76) system-atically
X and we can do a better job of (77) _______	(77) predicting
Y with knowledge of X than without it. The	
correlation (78) _______ expresses quantitatively	(78) coefficient
the (79) _______ and (80) _______	(79) magnitude (80) direction
of the relationship.The correlation coefficient	
can vary from (81) _______ to (82) _______.	(81) +1 (82) −1
The (83) _______ of the coefficient	(83) sign
tells us whether the relationship is positive or	
negative.The (84) _______ part of the coefficient	(84) numerical
tells us the magnitude of the correlation. The	
(85) _______ the absolute value of	(85) higher
the correlation coefficient, the stronger the	
relationship. A correlation of(86) _______ or	(86) +1

(87) _______ means the relationship is perfect,	(87) −1
and a correlation of (88) _______ means there is	(88) 0
(89) _______ relationship between the	(89) no
two variables. Imperfect relationships have	
magnitudes varying between	
(90) _______ and (91) _______. They will be	(90) 0 (91) 1
(92) _______ or (93) _______ depending	(92) plus (93) minus
on the direction of the relationship. In a positive	
relationship, as *X* (94) _______, *Y* (95) _______.	(94) increases (95) increases
In a (96) _______ relationship, as *X* increases,	(96) negative
Y decreases. In the case where there is a	
zero regression line for predicting	
Y is (97) _______.	(97) horizontal
Pearson *r* is defined as a measure of the	
extent to which paired scores occupy the	
same (98) _______ within their own	(98) position

distribution. When each pair of scores has the	
same (99) _______ value, Pearson *r* equals	(99) *z*
(100) _______ Standard scores (z scores)	(100) 1
are used in determining r to allow measurement	
of the relationship between the two variables	
which is(101) _______ of differences in	(101) independ-ent
(102) _______ and (103) _______.	(102) scaling (103) units

To calculate the value of *r* from raw scores, we can utilize the following equation:

$$r = \frac{(104)\_\_\_\_ - \frac{((105)\_\_\_\_)((106)\_\_\_\_)}{(107)\_\_\_\_}}{\sqrt{\left[(108)\_\_\_\_ - \frac{(109)\_\_\_\_}{(110)\_\_\_\_}\right]\left[(111)\_\_\_\_ - \frac{(112)\_\_\_\_}{(113)\_\_\_\_}\right]}}$$

(104) ΣXY
(105) ΣX
(106) ΣY
(107) N
(108) ΣX^2
(109) $(\Sigma X)^2$
(110) N
(111) ΣY^2
(112) $(\Sigma Y)^2$
(113) N

In predicting *Y* from *X* the (114) _______	(114) total
variability of *Y* can be divided into two parts;	
the variability of (115) _______ errors and the	(115) prediction

variability of (116) _______ accounted for	(116) *Y*
by (117) _______.	(117) *X*
The total variability is symbolized by	
(118) _______. The variability of prediction	(118) $\Sigma(Y_i - \overline{Y})^2$
errors is symbolized by (119) _______	(119) $\Sigma(Y_i - Y')^2$
The variability of *Y* accounted for by *X* is	
symbolized by (120) _______. As the	(120) $\Sigma\ (Y' - \overline{Y})^2$
relationship gets stronger the variability of	
prediction errors gets (121) _______ and the	(121) smaller
variability of *Y* accounted for by *X* gets	
(122) _______. Therefore, the proportion	(122) larger
of the total variability of the *Y* scores accounted	
for by *X* is another measure of the strength of the	
(123) _______. A conceptual formula for *r* is:	(123) relation-ship
$r = \dfrac{\sum((124)\_\_\_\_ - (125)\_\_\_\_)^2}{\sum((126)\_\_\_\_ - (127)\_\_\_\_)^2}$	(124) Y' (125) $\overline{Y}$ (126) Y_i (127) $\overline{Y}$

Squaring both sides of the above equation	
gives us useful information about how important	
X is in explaining the (128) _______ of *Y*.	(128) variability
Hence, r^2 = the (129) _______ of the total	(129) proportion
variability of (130) _______ accounted for	(130) *Y*
by (131) _______.	(131) *X*
Besides Pearson *r*, there are other	
correlation coefficients. When the data is	
curvilinear, one can use the correlation	
coefficient (132) _______. Pearson *r* is used	(132) eta
when data are measured on an (133) _______	(133) interva1
or (134) _______ scale. If one variable is at	(134) ratio
least (135) _______ and the other	(135) interval
(136) _______, the biserial correlation	(136) dichoto-mous
coefficient is used. If both variables are	
dichotomous, then (137) _______ is used.	(137) phi
Perhaps the second most often used	
correlation coefficient is the	

(138) _______ rank order coefficient, | (138) Spearman

symbolized by (139) _______. The equation | (139) r_S

for r_S is:

$$r_s = 1 - \frac{6\sum D_i^2}{(140)\ \_\_\_\_\_ - N}$$ | (140) N^3

where

D_i = the (141) _______ between the ith | (141) difference

pair of ranks

N = the number of (142) _______ of ranks. | (142) pairs

In doing correlational studies there are

certain things to keep in mind. If there is

not sufficient (143) _______ of both | (143) range

variables, the correlation may be artificially

low. Also, it is tempting to infer that correlation

means causation. This inference

would be unjustified without further

(144) _______. | (144) experimentation

There are four possible explanations for correlation. First, the correlation between *X* and *Y* may be (145) _______. Second, *X* may cause *Y* or, third, *Y* may cause (146) _______. Finally, a (147) _______ variable may cause the correlation between *X* and *Y*.

(145) spurious

(146) *X*

(147) third

EXERCISES

1. Using the following table, construct a scatter plot of the pairs of values for height and weight. Calculate the correlation between height and weight.

	Height		Weight		
Person	**X**	X^2	**Y**	Y^2	**XY**
1	73		185		
2	68		150		
3	61		125		
4	65		130		
5	70		162		
6	69		170		

2. The phone rings and the President is calling you. He says, "I hear you are studying correlation. Can you please tell me how much of the variability of my popularity rating is explained by the inflation rate?" He kindly supplies you with the following data:

Inflation rate	10	12	8	6	14
Popularity rating	70	75	85	84	48

3. An ornithologist wants to know how much of a relationship there is between the weight gain of the female flicker during the spring and the number of eggs laid. What is the Pearson r for the data the ornithologist gathered?

Bird	1	2	3	4	5	6
Weight Gain (oz.)	2	6	6	3	10	7
Number of Baby Flickers	0	1	3	3	6	5

4. Using the data of Problem 3, how much of the variability in the number of eggs is accounted for by knowing the amount of weight gained?

5. A teacher was interested in knowing if leadership ability was correlated with attractiveness in third graders. The teacher ranked a group of students for leadership ability and had another teacher rank the same children on attractiveness.

   a. What kind of scales are involved in this problem?
   b. What correlation coefficient is appropriate for this type of data?
   c. Using the following data and table as an aid, what is the value for r_S for these data?

Student	Leadership rank	Attractiveness rank	D^2
A	1	2	
B	2	1	
C	3	3	
D	4	7	
E	5	4	
F	6	5	
G	7	6	

6. Calculate the Pearson *r* for the following set of data:

X	1	2	3	4
Y	10	13	14	16

7. Using the following set of data:

X	1	2	3	4
Y	16	14	13	10

   a. Calculate the value of Pearson *r*.
   b. What is the difference between the data of problem 6 and problem 7? How does it relate to Pearson *r*?

8. Given the following set of data for six subjects:

Subject No.	1	2	3	4	5	6
X	6	9	7	7	5	15
Y	9	7	10	8	6	17

a. Construct a scatter plot of the data.
b. Calculate Pearson *r* for the first 5 subjects.
c. Next, add the data of subject 6, and recalculate Pearson *r* for all six subjects.
d. Explain the difference between the values of *r* for part **b** and part **c**.

9. Given the following set of data for 8 subjects:

Subject No	1	2	3	4	5	6	7	8	9	10
X	5	8	6	10	9	7	6	7	9	10
Y	7	9	7	12	8	8	9	10	11	10

a. Construct a scatter plot of the data.
b. Calculate Pearson *r*.
c. Remove the data for subjects 1, 3, 4, and 10. Recalculate *r*.
d. Explain the difference in the values of *r* obtained for part **b** and part **c**.

10. A study is conducted to determine the reliability of two judges in assessing musical performance. The judges are asked to rate 8 musical contestants on a twenty point scale. The higher the rating, the greater is the assessed musical talent. The following data are obtained:

Contestant No	1	2	3	4	5	6	7	8
Judge A	18	16	12	10	17	15	13	14
Judge B	16	13	14	7	18	11	9	17

Assuming the data are only of ordinal scaling, use the appropriate correlation coefficient to assess how alike the judges are in their ratings.

TRUE-FALSE QUESTIONS

T F 1. The easiest way to determine if a relationship is linear is to calculate the regression line.

T F 2. In a linear relationship all the points must fall on a straight line.

T F 3. In a perfect linear relationship all the points must fall on a straight line.

T F 4. The slope of a line is a measure of its rate of change.

T F 5. In a straight line the slope approaches zero as the line comes near the point X, Y.

T F 6. In an inverse relationship as one variable gets larger the other variable gets smaller.

T F 7. A correlation coefficient expresses the direction but not the magnitude of a relationship.

T F 8. Both Pearson r and Spearman rho can range from −1.00 to +1.00.

T F 9. The value of r obtained by calculating the correlation between X and Y is the same as the correlation between Y and X.

T F 10. If scores are z scores and if r equals 1 then z_X will always equal z_Y.

T F 11. One reason for calculating r from z scores is to make r independent of units and scaling.

T F 12. The coefficient of determination equals the proportion of variability accounted for by the relationship between the variables.

T F 13. The coefficient of determination equals $\sqrt{r}$.

T F 14. Since r is so widely used, it is appropriate to calculate r for nonlinear data.

T F 15. The formula for rho is actually just the formula for Pearson's r simplified to apply to lower order scaling.

T F 16. If the value of rho for a set of ordinal data equaled 0.68, the value of r for the same data would be 0.68.

T F 17. If the value of r on ratio scale raw data were 0.87 and the pairs of numbers were converted to ordinal data and r calculated for the ordinal data, r would equal 0.87.

T F 18. Restricting the range of either X or Y will generally lower the correlation between the variables.

T F 19. If one calculates r for a set of data and r equals 0.84, one can be certain that the relationship between the variables is not spurious.

T F 20. The correlation between two variables when $N = 2$ will always be perfect.

T F 21. The correlation coefficient when $N = 2$ is meaningless.

T F 22. If $r = -1.00$, the relationship is imperfect.

T F 23. If one calculates r for a set of numbers and then adds a constant to each value of one of the variables, the correlation will change.

T F 24. If the standard deviation of one of the variables equals zero, r cannot be calculated.

T F 25. The correlation coefficient r is a descriptive statistic.

SELF-QUIZ

1. Which of the following values of r represents the strongest degree of relationship between two variables?

   a. 0.55
   b. 0.00
   c. 0.78
   d. –0.80

2. What is the slope for the points $X_1 = 30$, $Y_1 = 50$ and $X_2 = 25$ and $Y_2 = 40$?

   a. 2.00
   b. 0.50
   c. –2.00
   d. –0.50

3. In order for the correlation coefficient to be negative, which of the following must be true?

    a. $\Sigma XY > (\Sigma X)(\Sigma Y)/N$
    b. $\Sigma XY < (\Sigma X)(\Sigma Y)/N$
    c. $\Sigma XY = (\Sigma X)(\Sigma Y)/N$
    d. ΣXY must be zero

4. If two variables are ratio scaled and the relationship is linear, what type of correlation coefficient is most appropriate?

    a. Pearson
    b. Spearman
    c. eta
    d. phi

5. Correlation means causation.

    a. true
    b. false

6. Causation means correlation.

    a. true
    b. false

7. Pearson *r* can be properly used on which of the following type(s) of relationships?

    a. linear
    b. curvilinear
    c. exponential
    d. all of the above

8. If one takes a sample of pairs of points over a narrow range of *X* or *Y* scores, what effect might this have on the value of *r*?

a. inflate r
b. have no effect on r
c. reduce r
d. cannot be determined

9. You have conducted a brilliant study which correlates IQ score with income and find a value of $r = 0.75$. At the end of the study you find out all the IQ scores were scored 10 points too high. What will the value of r be with the corrected data?

a. r will be increased
b. r will be decreased
c. r will remain the same
d. cannot be determined

10. If z_X equals z_Y for each pair of points, r will equal _______.

a. 0.00
b. −1.00
c. 1.00
d. 0.50

11. If one calculates r for raw scores, and then calculates r on the z scores of the same data, the value of r will _______.

a. stay the same
b. decrease
c. increase
d. equal 1.00

12. You have noticed that as people eat more ice cream they also have darker suntans. From this observation, you conclude _______.

a. eating ice cream causes people to tan darker

b. when one's skin tans it causes an urge to eat ice cream
c. the results were spurious
d. perhaps a third variable is responsible for the correlation
e. all of the above are possible

13. If 49% of the total variability of Y is accounted for by X, what is the value of r?

a. 0.49
b. 0.51
c. 0.70
d. 0.30

14. What is the value of r for the following relationship between height and weight?

Height	**60**	**64**	**65**	**68**
Weight	**103**	**122**	**137**	**132**

a. 0.87
b. 0.76
c. 0.93
d. 0.56

15. For the following X and Y scores, how much of the variability of Y is accounted for by knowledge of X? Assume a linear relationship.

X	**20**	**15**	**6**	**10**
Y	**6**	**5**	**4**	**0**

a. 68%
b. 34%
c. 58%
d. 27%

16. What is the value of the Spearman rank order correlation coefficient (rho) for the following pairs of ranks?

Rank on A	Rank on B
1	3
2	1
3	4
4	2
5	5

a. 0.40
b. 0.50
c. 0.60
d. 0.70

17. In order to properly use rho, the variables must be of at least _______ scaling.

a. nominal
b. ordinal
c. interval
d. ratio

ANSWERS

Exercises: **1.** 0.95 **2.** .75 or 75% (r^2) **3.** 0.79 **4.** .63 or 63% (r^2) **5a.** ordinal **5b.** Spearman rho **5c.** 0.75 **6.** 0.98 **7a.** –.98 **8b.** 0.11 **8c.** 0.88 **8d.** The data of

Subject 6 represents an extreme point with respect to the rest of the distribution. It has greatly increased the value of *r*. **9b.** 0.73 **9c.** 0.19 **9d.** Range restriction has lowered the value of *r* **10**. $r_s = 0.62$.

True-False: **1.** F **2.** F **3.** T **4.** T **5.** F **6.** T **7.** F **8.** T **9.** T **10.** T **11.** T **12.** T **13.** F **14.** F **15.** T **16.** T **17.** F **18.** T **19.** F **20.** T **21.** T **22.** F **23.** F **24.** T **25.** T.

Self-Quiz: **1.** d **2.** a **3.** b **4.** a **5.** b **6.** a **7.** a **8.** c **9.** c **10.** c **11.** a **12.** e **13.** c **14.** a **15.** b **16.** b **17.** b.

Chapter 7

Linear Regression

CHAPTER OUTLINE

I. Introduction.

A. Linear regression. This topic deals with predicting scores of one distribution using information known about scores of a second distribution. For example, one might predict your height if they knew your weight and the nature of your relationship between height and weight from a sample of other people.

B. Correlation. This refers to the magnitude and direction of the relationship between two variables.

II. Least-Squares Regression Line for Prediction.

A. Least-squares criterion. In an imperfect relationship no single straight line will hit all the points. We pick the line that will minimize the total errors of prediction, i.e., construct the one line that minimizes $\Sigma\,(Y - Y')^2$ where Y' is the predicted value of Y for any value of X.

B. Constructing the regression line of Y on X.

1. Equation.

$$Y' = b_Y X + a_Y$$

where

$$b_Y = \frac{\sum XY - \frac{(\sum X)(\sum Y)}{N}}{\sum X^2 - \frac{(\sum X)^2}{N}}$$

$$a_Y = \overline{Y} - b_Y \overline{X}$$

C. Use of regression equation. For a given value of X, simply plug that value in the equation and solve for Y' using the regression constants b_Y and a_Y.

D. Regression of X on Y. In some cases one wants to predict X given a value of Y. The equation of interest becomes:

$$X' = b_X Y + a_X$$

where

$$b_X = \frac{\sum XY - \frac{(\sum X)(\sum Y)}{N}}{\sum Y^2 - \frac{(\sum Y)^2}{N}}$$

$$a_X = \overline{X} - b_X \overline{Y}$$

III. Prediction Errors. When relationships between *X* and *Y* variables are imperfect, there will be prediction errors.

A. <u>Standard error of estimate ($s_{Y|X}$)</u>. Quantifying the magnitude of the error involves computing the standard error of estimate symbolized $s_{Y|X}$. The standard error is much like the standard deviation.

1. Definition. Gives a measure of the average deviation of the prediction errors about the regression line.
2. Equation for standard error of estimate.

$$s_{Y|X} = \sqrt{\frac{SS_Y - \frac{[\sum XY - (\sum X)(\sum Y)/N]^2}{SS_X}}{N-2}}$$

3. Interpretation. The larger the value of $s_{Y|X}$, the less confidence one has in the prediction of *Y* given *X*. The smaller the value of $s_{Y|X}$, the more likely the prediction will be accurate. If one constructed two parallel lines to the regression line at distances of $\pm 1 s_{Y|X}$, $\pm 2 s_{Y|X}$, and $\pm 3 s_{Y|X}$, one would find about 68%, 95%, and 99% of the scores would fall between the lines respectively.

B. <u>Other errors</u>. One must be careful of sources of errors in making predictions. There are two major considerations in making predictions.

1. Linearity. The original relationship needs to be linear for accurate prediction using linear regression.

2. Prediction in the range. Generally one uses a sample to generate the data for calculating the regression constants (b_Y and a_Y). Predictions of Y should be based on values of X within the range of the sample upon which the constants are based.

IV. Regression Constants and Pearson *r*

A. Regression coefficient. $\boldsymbol{b_Y = r(s_Y/s_X)}$

B. Regression coefficient. $\boldsymbol{b_X = r(s_X / s_Y)}$

C. Regression constant a_Y. found in the usual way

D. Slope of regression line for *z* scores. Equals *r*.

V. Multiple Regression

A. Extension of simple regression. Multiple regression is an extension of simple regression (single predictor) to situations that involve two or more predictor variables.

B. Prediction accuracy. Increases accuracy of prediction.

C. Equation for two predictor variables.

$$Y' = b_1X_1 + b_2X_2 + a$$

D. Multiple coefficient of determination, R^2. R^2 = multiple coefficient of determination = squared multiple correlation.

E. Equation of R^2 for two predictor variables.

$$R^2 = \frac{r_{YX_1}{}^2 + r_{YX_2}{}^2 - 2r_{YX_1}r_{YX_2}r_{X_1X_2}}{1 - r_{X_1X_2}{}^2}$$

CONCEPT REVIEW

It is often useful to use knowledge of one variable to predict a likely value on a second variable.

If there is a

(1) _______ between two variables, we can use knowledge of this relationship for (2) _______.

(1)relationship

(2) prediction

The name of this topic which covers this material for linear relationships is (3) _______ (4) _______. In an (5) _______

(3) linear
(4) regression
(5) imperfect

relationship, all of the points do not fall on the regression line. In an imperfect relationship one constructs the line which (6) _______ errors of (7) _______ according to a(8) _______.

(6) minimizes

(7) prediction
(8) least

(9) _______ criterion This is called the

(9) squares

(10) _______ (11) _______	(10) least (11) squares
(12) _______ (13) _______. The (14) _______	(12) regression (13) line (14) vertical
(15) _______ between the regression line and	(15) distance
each (16) _______ represents the(17) _______	(16) point (17) error
in prediction. (18) _______ equals the predicted	(18) Y'
Y value and (19) _______ equals the actual value	(19) Y
of Y. (20) _______ -(21) _______ equals the	(20) Y (21) Y'
error for each point. The least-squares	
regression line minimizes (22) _______.	(22) $\Sigma(Y - Y')^2$

Constructing the Regression Line

The terms (23) _______ and (24) _______ are called regression (25) _______	(23) b_Y (24) a_Y (25) constants
.The regression line for predicting Y given X is	
constructed *by* computing values for	(26) b_Y
(26) _______and (27) _______. The	(27) a_Y
computational formula for computing b_Y is:	

$$b_Y = \frac{(28)\_\_\_ - \dfrac{(29)\_\_\_(30)\_\_\_}{(31)\_\_\_}}{(32)\_\_\_ - \dfrac{(33)\_\_\_}{(34)\_\_\_}}$$

(28) ΣXY
(29) (ΣX)
(30) (ΣY)
(31) N
(32) ΣX^2
(33) $(\Sigma X)^2$
(34) N

N is equal to the number of (35) _______ (36) _______.

(35) paired
(36) scores

The a_Y regression constant is given by the equation:

a_Y = (37) _______ − (38) _______ (39) _______

(37) $\overline{Y}$
(38) b_Y
(39) $\overline{X}$

Since we need to know the value of b_Y to determine the a_Y constant, we first find (40) _______, then (41) _______. Once they are both found they are substituted into the (42) _______ equation.

(40) b_Y
(41) a_Y
(42) regression

The above regression constants are for the values of the regression line of (43) _______ on (44) _______

(43) Y
(44) X

It is sometimes of interest to predict X given Y.

This is called the regression line

of (45) _______ on	(45) X
(46) _______. The linear	(46) Y
regression equation for predicting X given Y is:	
(47) _____ = (48) _____ times	(47) X
	(48) b_X
(49) _____ + a_X	(49) Y
This regression line is constructed by calculating	
values for (50) _______ and (51) _______. The	(50) b_X
computational formula for b_X is:	(51) a_X
$$b_X = \frac{(52)\_\_\_ - \frac{(53)\_\_\_(54)\_\_\_}{(55)\_\_\_}}{(56)\_\_\_ - \frac{(57)\_\_\_}{(58)\_\_\_}}$$	(52) ΣXY (53) (ΣX) (54) (ΣY) (55) N (56) ΣY^2 (57) $(\Sigma Y)^2$ (58) N
The equation for a_X is:	
a_X = (59) _______ − (60) _______	(59) $\overline{X}$
	(60) b_X
(61) _______	(61) $\overline{Y}$

The regression line of Y on X will equal the

regression line of X on Y

only when the relationship is (62) _______.	(62) perfect

Measuring Prediction Errors

Quantifying prediction errors involves computing the	
(63) _______ (64) _______ of	(63) standard (64) error
(65) _______ that is symbolized by (66) _______.	(65) estimate (66) $s_{Y\|X}$
The standard error of estimate gives a measure of the	
(67) _______ deviation of the (68) _______	(67) average (68) prediction
errors about the (69) _______ (70) _______.	(69) regression (70) line
The conceptual formula for $s_{Y\|X}$ is:	
$s_{Y\|X}$ = (71) _____ /(72) _____	(71) $\Sigma(Y - Y')^2$ (72) $N - 2$
The computational formula is:	
$$s_{Y\|X} = \frac{(73)\_\_\_\_\_ - \frac{(74)\_\_\_\_\_}{(75)\_\_\_\_\_}}{(76)\_\_\_\_\_ - (77)\_\_\_\_\_}$$	(73) SS_Y (74) $[\Sigma XY - (\Sigma X)(\Sigma Y)/N]^2$ (75) SS_X (76) N (77) (2)
for predicting (78) _______	(78) Y

given (79) _______. The standard error of — (79) X

estimate is computed over all (80) _______ — (80) Y

scores. For it to be meaningful one assumes that the

(81) _______ of Y remains (82) _______ as — (81) variability (82) constant

one goes from one (83) _______ score to the next. — (83) X

This assumption is called the assumption of

(84) _______. — (84) homosce

In general dasticity one would expect to find

(85) _______ of the points to fall within — (85) 68%

$\pm 1\, s_{Y|X}$ of the regression line, 95% of

the points to fall within ±(86) _______ — (86) 2

$s_{Y|X}$, and (87) _______ % to fall within ± — (87) 99

(88) _______. — (88) $3\, s_{Y|X}$

In general it is appropriate to use linear

regression to predict values only if the

(89) _______ is (90) _______. It is also — (89) relationship (90) linear

basic (91) _______ group be important that the	(91) computation or sample
representative of the (92) _______ group. In other	(92) prediction
words, the data collected to compute the	
(93) ______ (94) _______ should be a	(93) regression
	(94) constants
(95) _______ sample from the	(95) random
(96) _______ of interest.	(96) population
Finally, the linear regression equation is properly	
used just for the (97) _______ of the variable	(97) range
upon which it is based. This is because we do	
not know if data outside the range of our	
sample continues to be a (98) _______	(98) linear
(99) _______.	(99) relationship
(100) _____ and	(100) b_Y
(101) _____ are the slopes of the regression	(101) b_X
lines when the scores are plotted as raw scores.	
Pearson (102) _____ is the slope of the	(102) r
regression line when the scores are plotted	

as (103) _____ | (103) z

scores. Because they are both slopes

of regression lines, (104) _____ and | (*104*) r

(105) _____ are related by an equation. | (105) b_Y

The equation is:

$$b_Y = (106)\_\_\_\_\_ \times \frac{(107)\_\_\_\_\_}{(108)\_\_\_\_\_}$$

(106) r
(107) s_Y
(108) s_X

For the slope of the regression line of X upon Y, the equation is:

$$b_X = (109)\_\_\_\_\_ \times \frac{(110)\_\_\_\_\_}{(111)\_\_\_\_\_}$$

(109) r
(110) s_X
(111) s_Y

Simple regression refers to

using (112) _______ predictor variable. | (112) one

Multiple regression refers to using

(113) _______ | (113) two or more

predictor variables. Using (114) _______ | (114) two or more

predictor variables

often increases the accuracy of prediction. Of

course, this depends on how much

(115) _______variation the additional predictor variables can account for. (115) unexplained

(116)_______ tells us how much variation is accounted for by all of the predictor variables. (116) R^2

EXERCISES

1. *X* represents aptitude test scores and *Y* represents grade point average in college. If the least-squares regression line for the relationship between these two variables is $Y' = .005X + 1.2$, what GPA would you predict for people who scored each of the following scores on the aptitude test?

   a. 159
   b. 300
   c. 500
   d. 550

2. Draw a graph of aptitude test score versus grade point average and construct the regression line for the line $Y' = .005X + 1.2$.

3. A professor wanted to predict final exam scores from midterm exam scores. He used data from several different professors teaching the same class. He obtained the following data:

Midterm Scores:	**83,**	**62,**	**72,**	**85,**	**85**
Final Exam Scores:	**89,**	**58,**	**70,**	**92,**	**84**

What are the values for each of the following?

a. ΣX.
b. ΣX^2.
c. N.
d. ΣY.
e. ΣY^2.
f. $(\Sigma X)^2$.
g. $(\Sigma Y)^2$.
h. ΣXY.
i. b_Y.
j. a_Y.
k. If the professor's class score on the midterm were 77.4, what score would you predict the class would receive on the final exam?
l. What is the value of $s_{Y|X}$?

4. A hospital administrator wanted to predict the number of patients her hospital would admit in 1990. The following data were obtained from past records:

Year	**1960**	**1965**	**1970**	**1975**	**1980**
Number of Admissions	**812**	**983**	**1127**	**904**	**1768**

a. What would the best prediction be for the number of admissions expected in 1990?
b. What serious caution should the administrator be aware of when making her prediction?

5. A psychologist wanted to use a locus of control test to predict scores on a depression scale. The following data were summarized for the relationship between the locus of control and depression scale:

$\Sigma X = 62$, $\Sigma X^2 = 1022$, $\Sigma Y = 70$, $\Sigma Y^2 = 1234$, $\Sigma XY = 1107$; $N = 4$

a. What is the value of b_Y?
b. What is the value of a_Y?
c. What would the psychologist predict for the score on the depression scale if a client scored an 18 on the locus of control scale?

6. Consider the following set of data points:

X	2	4	8	14	20	23	25
Y	2	6	14	20	12	9	7

a. Construct a scatter plot of the points.
b. Is it appropriate to use a least-squares 1inear regression line to predict *Y* from *X* in this case? Why or why not?

7. Consider the following set of points for variable *X* and variable *Y*:

X	21	29	33	40	50
Y	34	36	42	45	58

Assume the relationship is linear in answering the following questions.

a. What is the value of the regression constant b_X for predicting *X* given *Y*?
b. What is the value of the regression constant a_X for predicting *X* given *Y*?
c. What value of *X* would you predict for a value of $Y = 37$?
d. What value of *X* would you predict for a value of $Y = 55$?

8. What is the linear regression equation for predicting *Y* given *X* for the following pairs of scores:

X	9	15	25	27	42	50	30
Y	14	11	5	5	0	–8	1

9. Given the following data, what are the values of b_Y and a_Y?

$$s_X = 12 \quad s_Y = 14 \quad \Sigma X = 57 \quad \Sigma Y = 83 \quad N = 10 \quad r = .843$$

TRUE-FALSE QUESTIONS

T F 1. In regression analysis we are only concerned with perfect as opposed to imperfect relationships.

T F 2. If we minimize $\Sigma (Y - Y')^2$, we will minimize the total error of prediction.

T F 3. The value a_Y is the X axis intercept for minimizing errors in Y.

T F 4. Generally, one can use the same regression equation for predicting Y given X as for X given Y.

T F 5. If the relationship between two variables is perfect the standard error of estimate equals 0.

T F 6. If the standard error of estimate for relationship 1 equals 5.26 and for relationship 2 it equals 8.01 then we can reasonably infer that relationship 2 is less perfect than relationship 1.

T F 7. It is impossible to have a negative value for the standard error of estimate.

T F 8. In general one is less confident in predictions of *Y* when the value of *X* used for the prediction is outside the range of the original data used to construct the regression line.

T F 9. If the regression line is parallel to the *X* axis then the slope of the regression line equals 0.

T F 10. The regression line will always go through the point $\overline{X}, \overline{Y}$. (Try it and see.)

T F 11. If *X* and *Y* are plotted as standard (*z*) scores, then *r* equals the slope of the resulting regression line.

T F 12. If $s_Y = s_X$ then $r = b_Y$.

T F 13. Using a second predictor variable always increases the accuracy of prediction.

SELF-QUIZ

1. If $s_{Y|X}$ = 0.0 the relationship between the variables is _______.

   a. perfect
   b. imperfect
   c. curvilinear
   d. unknown

2. $\Sigma (Y - Y')$ equals _______.

   a. 0
   b. 1

c. cannot be determined from information given
d. who cares

3. $\Sigma (Y - Y')^2$ represents _______.

a. the standard deviation
b. the variance
c. the standard error of estimate
d. the total error of prediction

4. In a particular relationship $N = 80$. How many points would you expect on the average to find within $\pm 1\, s_{Y|X}$ of the regression line?

a. 40
b. 80
c. 54
d. 0

5. What would you predict for the value of Y for the point where the value of X is $\overline{X}$?

a. cannot be determined from information given
b. 0
c. 1
d. $\overline{Y}$

6. If the value of $s_{Y|X} = 4.00$ for relationship *A* and $s_{Y|X} = 5.25$ for relationship *B*, in which relationship would you have the most confidence in a particular prediction?

a. *A*
b. *B*
c. it makes no difference
d. cannot be determined from information given

7. If b_Y is negative, higher values of X are associated with _______.

   a. lower values of X'
   b. higher values of Y
   c. higher values of $(Y - Y')$
   d. lower values of Y

8. Which of the following statement(s) is (are) an important consideration(s) in applying linear regression techniques?

   a. the relationship should be linear
   b. both variables must be measured in the same units
   c. predictions for Y should be within the range of the X variable in the sample
   d. a and c

9. In the regression equation $Y' = X$, the Y-intercept is _______.

   a. $\overline{X}$
   b. $\overline{Y}$
   c. 0
   d. 1

10. If the value for a_Y is negative, the relationship between X and Y is _______.

    a. positive
    b. negative
    c. inverse
    d. cannot be determined from information given

11. If $b_Y = 0$, the regression line is _______.

    a. horizontal
    b. vertical
    c. undefined
    d. at a 45° angle to the X axis

12. The least-squares regression line minimizes _______.

    a. s
    b. $s_{Y|X}$
    c. $\Sigma (Y - \overline{Y})^2$
    d. $\Sigma (Y - Y')^2$
    e. b and d

13. The points (0,5) and (5,10) fall on the regression line for a perfect positive linear relationship. What is the regression equation for this relationship?

    a. $Y' = X + 5$
    b. $Y' = 5X$
    c. $Y' = 5X + 10$
    d. Cannot be determined from information given.

14. For the following points what would you predict to be the value of Y' when $X = 19$? Assume a linear relationship.

X	6	12	30	40
Y	10	14	20	27

    a. 16.35
    b. 24.69
    c. 22.00
    d. 17.75

15. If $N = 8$, $\Sigma X = 160$, $\Sigma X^2 = 4656$, $\Sigma Y = 79$, $\Sigma Y^2 = 1309$, and $\Sigma XY = 2430$, what is the value of b_Y?

    a. 0.9217
    b. −1.8010
    c. 0.5838
    d. 0.7922

16. If X and Y are transformed into z scores, and the slope of the regression line is −0.80, what is the value of the correlation coefficient?

    a. −0.80
    b. 0.80
    c. 0.40
    d. −0.40

17. If the regression equation for a set of data is $Y' = 2.650X + 11.250$ then the value of Y' for X = 33 is _______.

    a. 87.45
    b. 371.25
    c. 98.70
    d. 76.20

18. If $\overline{X} = 57.2$, $\overline{Y} = 84.6$, and $b_Y = .37$, the value of $a_Y =$ _______.

    a. 141.80
    b. -25.90
    c. 63.44
    d. 27.40

19. If the regression line for predicting X given Y were $X' = 103Y + 26.2$, what would the value of X' be if $Y = 0.2$?

    a. 129.2
    b. 25.8
    c. 5.2
    d. 46.8

20. If $s_Y = s_X = 1$ and the value of $b_Y = .6$, what will the value of r be?

    a. 0.36
    b. 0.60
    c. 1.00
    d. 0.00

21. When using more than one predictor variable, _______ tells us the proportion of variance accounted for *by* the predictor variables.

    a. r
    b. SS_X
    c. SS_Y
    d. R^2

ANSWERS

Exercises: 1a. 2.00 **1b.** 2.70 **1c.** 3.70 **1d.** 3.95

2.

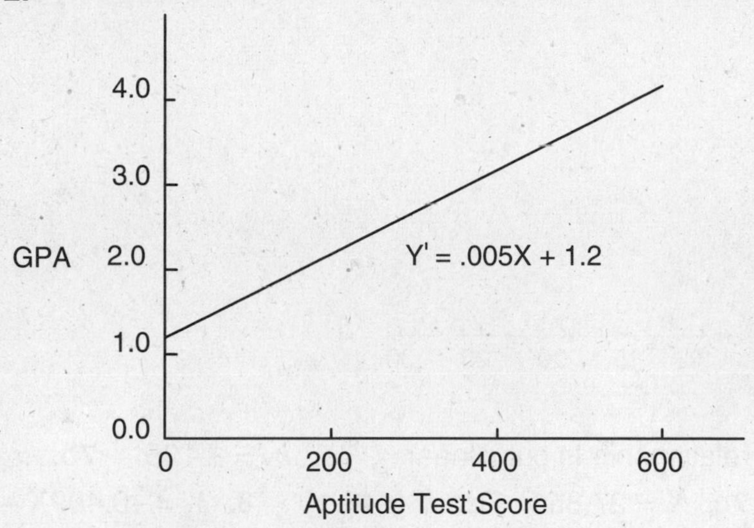

3a. 387 **3b.** 30,367 **3c.** 5 **3d.** 393 **3e.** 31,705 **3f.** 149,769 **3g.** 154,449 **3h.** 30,983 **3i.** 1.367 **3j.** -27.198 **3k.** 78.6 **3l.** 3.285 **4a.** 1852 **4b.** When predicting beyond the range of the sample data, one cannot be sure the nature of the relationship will remain the same. **5a.** 0.3607 **5b.** 11.91 **5c.** 18.4

6a.

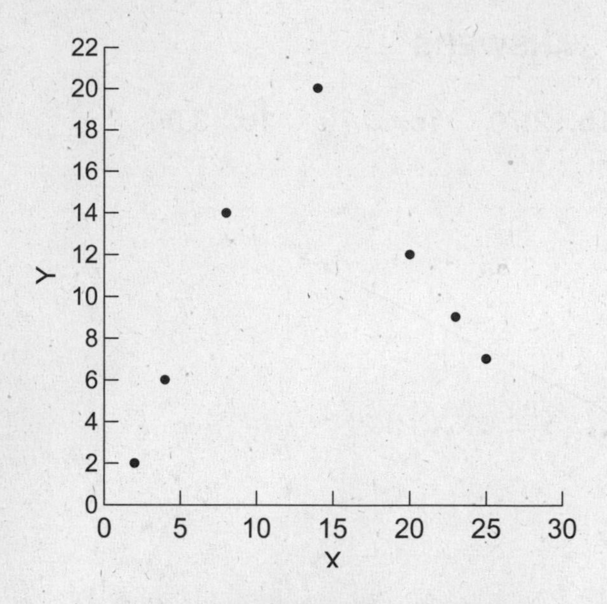

6b. No, the relationship is curvilinear **7a.** $b_X = 1.125$ **7b.** $a_X = -13.775$ **7c.** $X' = 27.85$ **7d.** $X' = 48.10$ **8.** $Y' = -0.499X + 18.126$ **9.** $b_Y = 0.984$ $a_Y = 2.691$.

True-False: **1.** F **2.** T **3.** F **4.** F **5.** T **6.** T **7.** T **8.** T **9.** T **10.** T **11.** T **12.** T **13.** F.

Self-Quiz: **1.** a **2.** a **3.** d **4.** c **5.** d **6.** a **7.** d **8.** d **9.** c **10.** d **11.** a **12.** e **13.** a **14.** a **15.** c **16.** a **17.** c **18.** c **19.** d **20.** b **21.** d.

Chapter 8

Random Sampling and Probability

CHAPTER OUTLINE

I. **Inferential Statistics.** Uses the sample scores to make a statement about a characteristic of the population.

   A. Hypothesis testing. Data collected in an experiment in an attempt to validate some hypothesis involving a population.

   B. Parameter estimation. The experimenter is interested in determining magnitude of a population characteristic; e.g., population mean.

II. **Methodology of Inferential Statistics**

   A. Random sampling. A random sample is defined as a sample which has been selected from the population by a process which assures that each possible sample of a given size has an equal chance of being selected, and all the members of the population have an equal chance of being selected into the sample.

B. Reasons for random sampling.

1. Required in order to apply laws of probability to sample.
2. It helps assure that the sample is representative of the population.

C. Random number table. Use of random number table is one method of assuring random sampling.

D. Types of sampling.

1. Sampling with replacement. A method of sampling in which each member of the population selected for the sample is returned to the population before the next member is selected.
2. Sampling without replacement. A method of sampling in which the members of the sample are not returned to the population prior to selecting subsequent members.

III. Probability

A. Classical or *a priori* view of probability. That which can be deduced from reason alone. No actual data required.

$$p(A) = \frac{\text{the number of events classifiable as } A}{\text{the total number of possible events}}$$

where $p(A)$ is read "The probability of occurrence of event A."

B. Empirical or *a posteriori* view of probability. Meaning after-the-fact or after some data has been collected.

$$p(A) = \frac{\textbf{the number of times } A \textbf{ has occurred}}{\textbf{the total number of occurrences}}$$

C. Probability values.

1. Range from 0.00 to 1.00 (i.e., from an event is certain not to occur to an event is certain to occur).
2. Generally expressed as fraction or decimal.

D. Computing probability.

1. Addition rule. This rule is concerned with determining the probability of occurrence of any of several possible events. Rule for two events:

   The probability of occurrence of *A* or *B* is equal to the probability of occurrence of *A* plus the probability of occurrence of *B* minus the probability of occurrence of both *A* and *B*, or

   $$p(A \text{ or } B) = p(A) + p(B) - p(A \text{ and } B)$$

2. Multiplication rule. This rule is concerned with the joint or successive occurrence of several events. Rule for two events:

   The probability of occurrence of both *A* and *B* is equal to the probability of occurrence of *A* times the probability of occurrence of *B* given *A* has occurred, or

   $$p(A \text{ and } B) = p(A)p(B|A)$$

   Both rules can be extended to more than two events.

E. Additional concepts.

1. Mutually exclusive events. Two events are mutually exclusive if they both cannot occur together (e.g., a coin coming up both heads and tails on one flip).
2. Exhaustive events. A set of events is exhaustive if the set includes all of the possible outcomes (e.g., for the flip of a coin, head and tail is the exhaustive set of possible outcomes).
3. Mutually exclusive and exhaustive events. When a set of events is both mutually exclusive and exhaustive the sum of the individual probabilities of each event in the set must equal 1.00. Thus, under these conditions:

$$p(A) + p(B) + p(C) + \cdots + p(Z) = 1.00$$

4. When there are two events and they are mutually exclusive and exhaustive, then:

$$P + Q = 1.00$$

IV. More on the Multiplication Rule

A. Mutually exclusive events. If A and B are mutually exclusive, then:

$$p(A \text{ and } B) = 0$$

B. Independent events. Two events are independent if the occurrence of one has no effect on the probability of occurrence of the other. In this case the rule becomes:

$$p(A \text{ and } B) = p(A)p(B|A) = p(A)p(B)$$

C. Dependent events. Two events are dependent if the probability of occurrence of B is affected by the occurrence of A. In this case the rule becomes:

$$p(A \text{ and } B) = p(A)p(B|A)$$

D. Multiplication and addition rules. Multiplication and addition rules can be combined to solve problems.

V. Probability and Continuous Variables

A. Equation.

$$p(A) = \frac{\text{the area under the curve corresponding to } A}{\text{the total area under the curve}}$$

B. Solution.

1. Convert raw score to its transformed value (its standard score).
2. Look up area in Table A.

CONCEPT REVIEW

A basic aim of (1) _______ statistics is to	(1) inferential
use (2) _______ scores to make a statement	(2) sample
bout a characteristic of the population. The	
characteristic may be either (3) _______	(3) hypothesis
testing or (4) _______ (5) _______. Essential	(4) parameter (5) estimation
to the methodology of inferential statistics are	
(6) _______ sampling and (7) _______ A random	(6) random (7) probability

sample is one which has been selected from the	
(8) _______ by a process which assures that	(8) population
each possible (9) _______ of a given size has	(9) sample
an (10) _______ chance of being selected and	(10) equal
that all the (11) _______ of the population have	(11) members
an(12) _______ chance of being selected into	(12) equal
the sample. Sampling should be random for two	
reasons. First, to apply the (13) _______ of	(13) laws
(14) _______ sampling must be random. Second,	(14) probability
in order to (15) _______ it is necessary that the	(15) generalize
sample be (16) _______ of the population.	(16) representtative
If a sample is (17) _______ it may lead to an	(17) biased
erroneous conclusion. One way to help assure	
random sampling is by the use of a table of	
(18) _______ (19) _______.	(18) random (19) numbers
In probability theory there can be two kinds of	
sampling. One method is called sampling with	
(20) _______. In this case each member of the	(20) replacement

population selected for the sample is (21) _______	(21) returned
to the population (22) _______ the next member	(22) before
is selected. The other method is sampling	
(23) _______ replacement. In this case the	(23) without
members of the sample are (24) _______	(24) not
(25) _______ to the population	(25) returned
(26) _______ selecting subsequent members.	(26) before
Probability can be approached in two ways,	
from an (27) _______ (28) _______ approach,	(27) *a* (28) *priori*
also called the (29) _______viewpoint or from	(29) classical
the (30) _______ (31) _______ or empirical	(30) *a* (31) *posteriori*
approach. The symbol (32) _______ means the	(32) $p(A)$
probability of the occurrence of event *A*. From	
the classical viewpoint, probability is defined:	
$$p(A) = \frac{\text{the (33) \_\_\_ of events (34) \_\_\_ as } A}{\text{the (35) \_\_\_ number of (36) \_\_\_ events}}$$	(33) number (34) classifiable (35) total (36) possible

From the empirical perspective, probability can be defined:

$$p(A) = \frac{\text{the number of (37)\_\_\_ } A \text{ has occurred}}{\text{the total number of (38) \_\_\_}}$$

(37) times
(38) occurrences

In this latter approach, actual (39) _______ must be collected. If enough data is collected the two probabilities will (40) _______ each other if chance alone is involved.

(39) data
(40) equal

Since probability is a (41) _______, the values can range from (42) _______ to (43) _______. If the probability equals 0.00, then the event is certain (44) _______ to occur. If the probability equals (45) _______, the event is certain to occur.

(41) proportion
(42) 0.00
(43) 1.00
(44) not
(45) 1.00

The addition rule states that the probability of occurrence of *A* (46) _______ *B* is equal to probability of occurrence of *A* (47) _______

(46) or
(47) plus

the probability of occurrence of *B* (48) _______	(48) minus
the probability of occurrence of (49) _______ *A*	(49) both
(50) _______ *B*. In equation form:	(50) and
$p(A \text{ or } B)$ = (51) _____ +	(51) $p(A)$
(52) _____ − (53) _____	(52) $p(B)$ (53) $p(A \text{ and } B)$
This rule is generally used when events are	
(54) _______ exclusive. This means that	(54) mutually
the occurrence of one event (55) _______	(55) precludes
the occurrence of the other. When two events	
exclusive are mutually then the addition rule	
becomes: $p(A \text{ or } B)$ = (56) _____ + (57) _____	(56) $p(A)$ (57) $p(B)$
This is true because $p(A \text{ and } B)$ = (58) _______.	(58) 0
A set of events is (59) _______ if the set includes	(59) exhaustive
all of the possible events. When events are both	
mutually exclusive and exhaustive, the	
(60) _______ of the individual probabilities of	(60) sum
each event in the set must equal (61) _______.	(61) 1.00

If we call one event P and the other Q and if	
P and Q are both mutually exclusive and	
exhaustive then this can be expressed as	
(62) _______ + (63) _______ = (64) _______	(62) P (63) Q (64) 1.00
The multiplication rule states the probability	
of the occurrence of A (65) _______ B is equal	(65) and
to the probability of occur rence of	
A (66) _______ the probability of B given A has	(66) times
occurred. In equation form:	
$p(A \text{ and } B)$ = (67) _______	(67) $p(A) \times p(B\|A)$
The term (68) _______ means the probability of	(68) $p(B\|A)$
occurrence of B given (69) _______ has occurred.	(69) A
When A and B are mutually (70) _______ then	(70) exclusive
$p(A \text{ and } B)$ = (71) _______.	(71) 0
Two events are considered (72) _______ if	(72) independent
the occurrence of one has (73) _______ effect	(73) no
on the probability of the other.In this case	

$p(B|A)$ = (74) _______. The multiplication rule (74) $p(B)$

then becomes: $p(A \text{ and } B)$ = (75) _______ (75) $p(A)p(B)$

The original formula used earlier, namely,

$p(A \text{ and } B) = p(A)p(B|A)$ is used when A and B are

(76) _______. This means the probability (76) dependent

of B is (77) _______ by the occurrence of A. (77) affected

One common example of a dependent event

occurs when one samples (78) _______ (78) without

replacement. For some problems we can use

both the multiplication and the addition rules

together.

So far we have been discussing (79) _______ (79) discrete

variables. When a variable is (80) _______ (80) continuous

probability can be defined as

$$p(A) = \frac{\text{the (81)\_\_\_ under the curve corresponding to } A}{\text{the (82)\_\_\_ area under the curve}}$$

(81) area
(82) total

Often these variables are normally distributed.

In that case we can find the area under the

curve by converting the (83) _______ score — (83) raw

to its (84) _______ value or (85) _______ — (84) z

score and then looking up the area in

Table A. We recall that — (85) z

this transformation is made by

$$z = \frac{(86)\_\_\_ - (87)\_\_\_}{(88)\_\_\_}$$

(86) X
(87) μ
(88) σ

EXERCISES

1. A gambler gives you a coin that he tells you is a "trick" coin. That is, the probability of getting a head does not equal the probability of getting a tail. To use the coin to your advantage you need to know the probability of obtaining a head; i.e., p(H).

   a. How can you determine p(H)?
   b. Assuming you tossed the coin 5000 times and got 1900 tails, what would you estimate p(H) to be?
   c. What method of determining probability is this called?

2. In a class of 30 children who cannot swim, the instructor randomly selects a child and then teaches the child to swim.

   a. What was the probability that any one particular child would have been selected?
   b. What is the probability that if the process is repeated a second time that a particular child will be selected?

c. What type of sampling technique is this?

3. The local psychological association reports that in a particular community there are 36 analytic psychologists, 57 behaviorists, 92 cognitive psychologists, 17 Gestalt psychologists and 42 of unknown theoretical background.
   a. If you randomly select a psychologist from a list containing all the above psychologists, what is the probability you will select either a known behaviorist or a known analytic psychologist? Assume sampling with replacement for all parts of this problem.
   b. What is the probability that you would select a psychologist of unknown background?
   c. What is the probability that you will select a psychologist with a known background?
   d. If you draw two psychologists, what is the probability that you will draw a known cognitive psychologist and a known Gestalt psychologist in that order?

4. A store that sells health foods, has 16 different types of flour. Four of the flours are white and the rest are brown. Assume sampling with replacement.

   a. Assuming random selection, what is the probability of selecting a brown or white flour?
   b. What is the probability of selecting a white and then a brown flour?
   c. What is the probability of selecting a brown and then a white flour?
   d. What is the probability of selecting three brown flours in a row?

5. Consider an ordinary deck of playing cards. What is the probability for the following:

   a. Drawing a red card or a king?

b. Drawing an ace, king, queen, jack, and 10 of the same suit in that order without replacement?
c. Drawing a face card (K, Q or J)?
d. Drawing 4 threes in four draws without replacement?
e. Drawing 4 threes in four draws with replacement?

6. A cattle breeder has 100 black cows of breed X, 50 brown cows of breed Y and 22 white cows of breed Y.

    a. What is the probability of randomly selecting a white cow and having it be a member of breed X?
    b. What kind of events are the events described in part a?
    c. What is the probability of randomly selecting a cow and having it be brown or of breed X?
    d. What is the probability of selecting at random and with replacement, first a brown cow, then a white cow, and finally a cow that is either of breed X or white?

7. Prior to taking an exam of 5 true or false items, assume (as ludicrous as this seems) that you did not have a chance to study. Assume further that you just have to guess on each item with the resulting probability of getting any item correct is .50. What is the probability for each of the following outcomes?

    a. That you get all 5 items correct?
    b. That you miss all 5 items?
    c. That you miss item 2?
    d. That you correctly answer the even number items?
    e. That you get both the first 2 items correct or fifth item correct?

8. If the average survival of a red cell in the human body is 120 days with a standard deviation of 8 days, what is the probability for each of the following events? Assume a normal distribution.

a. A red cell surviving more than 134 days?
b. A red cell that survives less than 110 days?
c. A red cell surviving between 108 and 122 days?

9. At the local racetrack 10 races are run each day with 10 different horses in each race. You bet on a horse in each race but select the horse by using a random number table.

   a. What is the probability you will choose a winner the first three races?
   b. What is the probability you will choose a winner in race 6?
   c. Does the fact that horse number 5 wins the third race affect the probability that any other number 5 will win during the day?

10. In an experiment on parapsychology a scientist wants to investigate how well a subject can correctly guess what symbol the experimenter is thinking of. The experimenter chooses from 4 different possible symbols. To avoid bias the experimenter uses a random number table to select a symbol to concentrate on for each trial. Assuming the experiment is not biased what are the probabilities for the following outcomes if chance alone is operating?

    a. A subject identifying the first 3 symbols in a row correctly?
    b. A subject missing the first 5 in a row?
    c. The subject guessing the nth symbol correctly

11. What is the probability that you can correctly guess the spelling of a three-letter sequence that you do not know, if the first letter is different from the second and third letters but the second and third letters may or may not be the same?

12. What is the probability that you can correctly guess a three-letter sequence if the first letter is a consonant and the second and third letters are the same vowel (a, e, i, o, u, or y)?

13. A normally distributed continuous variable has a value of μ = 60 and □ = 14. If one draws a score from the distribution, what is the probability that it will be:

    a. between 38 and 60?
    b. between 68 and 74?
    c. less than 60?
    d. less than 59?
    e. greater than 40.1?

14. In a box there are 10 slips of paper and each slip of paper has a number from 0 to 9 on it so that all the numbers appear once and only once.

    a. What is the probability that if you draw 3 slips of paper in a row with replacement the digits would be 1, 2, 3?
    b. If you draw 4 slips one-at-a-time with replacement, what is the probability that the number resulting from the four digits will be even?
    c. What is the probability that in one draw you will draw a number less than 4 or greater than 7?

15. What is the probability of throwing a pair of dice and having the sum of the dice equal 7 or 11?

16. Approximately how many times out of 100 could you throw a pair of dice 5 times and not come up with a total of 7 or 11?

TRUE-FALSE QUESTIONS

T F 1. Hypothesis testing is part of inferential statistics while parameter estimation is used in descriptive statistics.

T F 2. In order to generalize to the population a sample must be randomly selected.

T F 3. For a sample to be random, all the members must have an equal chance of being selected into the sample.

T F 4. To use a random number table properly one must begin on the top left hand column and read across.

T F 5. Sampling without replacement is always used in choosing subjects for an independent groups design experiment.

T F 6. An *a posteriori* approach to probability is never used because it is only an approximation of the true probability.

T F 7. Probability values range from −1.00 to +1.00.

T F 8. Events that occur only rarely have a probability equal to 0.00.

T F 9. Two events are considered mutually exclusive if the probability of one event does not influence the probability of a second event.

T F 10. When two events are dependent, then $p(A \text{ and } B) = p(A) + p(B)$.

T F 11. When events are mutually exclusive and exhaustive then the sum of the individual probabilities of each event in the set must equal 1.00.

T F 12. The addition rule is concerned with determining the probability of *A* or *B* while the multiplication rule is concerned with determining the probability *A* and *B*.

T F 13. When two events are mutually exclusive then $p(A \text{ and } B)$ must equal 1.00.

T F 14. Two events are independent if the occurrence of one has no effect on the probability of occurrence of the other.

T F 15. If A and B are independent then $p(B|A) = p(A)$ or $p(B)$.

T F 16. For dependent events, $p(A \text{ and } B) = p(A)p(B|A)$.

T F 17. The addition and multiplication rules can apply to any number of events.

T F 18. The multiplication and addition rules can be applied together in the same problem in order to calculate probabilities under some circumstances.

T F 19. There is no division rule for probability.

T F 20. To use probability and inference, data must be discrete.

T F 21. For continuous variables which are normally distributed, $p(A)$ equals the area under the curve corresponding to A divided by the total area under the curve.

T F 22. To calculate the probability of drawing three sevens in a row from a deck of cards involves the use of the multiplication rule and equals 4/52 + 3/51 + 2/50.

SELF-QUIZ

1. If a town of 7000 people has 4000 females in it, then the probability of randomly selecting 6 females in six draws (with replacement) equals _______.

   a. 0.0348
   b. 0.0571

c. 0.5714
d. 0.3429

2. If a stranger gives you a coin and you toss it 1,000,000 times and it lands on heads 600,000 times, what is p(Heads) for that coin?

a. 0.5000
b. 0.6000
c. 0.4000
d. 0.0000

3. The probability of randomly selecting a face card (K, Q, or J) or a spade in one draw equals _______.

a. 0.0192
b. 0.0577
c. 0.4808
d. 0.4231

4. The probability of drawing an ace followed by a king (without replacement) equals _______.

a. 0.0044
b. 0.0060
c. 0.0045
d. 0.0965

5. The probability of throwing two ones with a pair of dice equals _______.

a. 0.3600
b. 0.1667
c. 0.0278
d. 0.3333

6. If p(A or B) = p(A) + p(B) then A and B must be _______.

   a. dependent
   b. mutually exclusive
   c. overlapping
   d. continuous

7. If P + Q = 1.00 then P and Q must be _______.

   a. mutually exclusive
   b. exhaustive
   c. random
   d. a and b

8. If $\mu = 35.2$ and □ = 10, then p(X) for $X \leq 39$ equals _______. Assume random sampling.

   a. 0.3520
   b. 0.6200
   c. 0.1480
   d. 0.6480

9. If p(A and B) = 0, then A and B must be _______.

   a. independent
   b. mutually exclusive
   c. exhaustive
   d. unbiased

10. If p(A)p(B|A) = p(A)p(B), then A and B must be _______.

   a. independent
   b. mutually exclusive
   c. random
   d. exhaustive

11. If $p(A \text{ and } B) = p(A)p(B|A) \neq p(A)p(B)$, then A and B are _______.

    a. mutually exclusive
    b. random
    c. independent
    d. dependent

12. If $p(A) = 0.6$ and $p(B) = 0.5$, then $p(B|A)$ equals _______.

    a. 0.8333
    b. 0.3000
    c. 0.5000
    d. Cannot be determined from the information given

13. If $\mu = 400$ and □ = 100 the probability of selecting at random a score less than or equal to 370 equals _______.

    a. 0.1179
    b. 0.6179
    c. 0.3821
    d. 0.8821

14. If you have 15 red socks (individual, not pairs), 24 green socks, 17 blue socks, and 100 black socks, what is the probability you will reach in the drawer and randomly select a pair of green socks? (Assume sampling without replacement.)

    a. 0.3022
    b. 0.3077
    c. 0.0228
    d. 0.0227

15. If the probability of drawing a member of a population is not equal for all members, then the sample is said to be _______.

    a. random
    b. independent
    c. exhaustive
    d. biased

16. The probability of rolling an even number or a one on a throw of a single die equals _______.

    a. 0.6667
    b. 0.5000
    c. 0.0834
    d. 0.3333

17. If events are mutually exclusive they cannot be _______.

    a. independent
    b. exhaustive
    c. related
    d. all the above

18. The probability of correctly guessing a two digit number is _______.

    a. 0.1000
    b. 0.0100
    c. 0.2000
    d. 0.5000

19. When events A and B are mutually exclusive but not exhaustive, p(A or B) equals _______.

    a. 0.50
    b. 0.00
    c. 1.00
    d. Cannot be determined from the information given

20. The probability of correctly calling 4 tosses of an unbiased coin in a row equals _______.

    a. 0.0625
    b. 0.5000
    c. 0.1250
    d. 0.2658

ANSWERS

Exercises: 1a. Take the coin and flip it a large number of times and observe the occurrence of the event of interest, namely a head occurring and then apply the formula: p(H) = (the number of heads)/(the number of flips) **1b.** 3100/5000 = 0.6200 **1c.** Empirical or *a posteriori* method **2a.** 1/30 = 0.0333 **2b.** 1/29 = 0.0345 **2c.** sampling without replacement since once a child has learned to swim s/he cannot be returned to the population to be drawn again **3a.** 0.3811 **3b.** 0.1721 **3c.** 1 – 0.1721 = 0.8279 **3d.** 0.0263 **4a.** 1.00 **4b.** 0.1875 **4c.** 0.1875 **4d.** 0.4219 **5a.** 0.5385 **5b.** 4/52 x 1/51 x 1/50 x 1/49 x 1/48 = 0.00000001283 **5c.** 12/52 = 0.2308 **5d.** 4/52 x 3/51 x 2/50 x 1/49 = 0.000003694 **5e.** $(4/52)^4$ = 0.00003501 **6a.** 0 **6b.** mutually exclusive **6c.** 0.8721 **6d.** 50/172 x 22/172 x (100/172 + 22/172) = 0.0264 **7a.** $.5^5$ = 0.0312 **7b.** $.5^5$ = 0.0312 **7c.** 0.5000 **7d.** $.5^2$ = 0.2500 **7e.** .5 x .5 + .5 = 0.7500 **8a.** 0.0401 **8b.** 0.1056 **8c.** 0.5319 **9a.** $1/10^3$ = 0.0010 **9b.** 0.1 **9c.** no **10a.** $(1/4)^3$ = 0.0156 **10b.**

$(3/4)^5 = 0.2373$ **10c.** 1/4 = 0.2500 **11.** 1/26 x 1/25 x 1/25 = 0.00006154 **12.** 1/20 x 1/6 x 1/1 = 0.008333 **13a.** 0.4418 **13b.** 0.1256 **13c.** 0.5000 **13d.** .4721 **13e.** 0.9222 **14a.** 0.001 **14b.** 0.50 **14c.** 0.6000 **15.** 8/36 = 0.2222 **16.** approximately 28 on the average (0.2847 x 100).

True-False: **1.** F **2.** T **3.** T **4.** F **5.** T **6.** F **7.** F **8.** F **9.** F **10.** F **11.** T **12.** T **13.** F **14.** T **15.** F **16.** T **17.** T **18.** T **19.** T **20.** F **21.** T **22.** F.

Self-Quiz: **1.** a **2.** b **3.** d **4.** b **5.** c **6.** b **7.** d **8.** d **9.** b **10.** a **11.** d **12.** d **13.** c **14.** c **15.** d **16.** a **17.** a **18.** b **19.** d **20.** a.

Chapter 9

BINOMIAL DISTRIBUTION

CHAPTER OUTLINE

I. Binomial Distribution

A. Definition. This is a probability distribution that results when the following five conditions are met:

1. Series of *N* trials.
2. Only 2 possible outcomes on each trial.
3. Outcomes are mutually exclusive.
4. Each trial is independent of every other trial.
5. The probability of each possible outcome stays constant from trial to trial.

B. Information obtained.

1. Each possible outcome of the *N* trials.
2. Probability of getting each of the possible outcomes.

C. Example - "Flipping a Coin".

1. Each flip is a trial.
2. Only 2 possible outcomes.

3. Mutually exclusive because only a head or tail can occur.
4. Independent; i.e., outcome of one flip doesn't affect any other flip.
5. Probability of a head or tail stays constant from trial to trial.

II. Binomial Expansion.

A. Avoids enumeration. Avoids having to enumerate outcomes of binomial events.

B. Formula.

$$(P + Q)^N$$

where $P =$ probability of one of the two possible outcomes on a trial

$Q =$ probability of the other possible outcome = $(1 - P)$

$N =$ number of trials

C. Interpretation. The letters in each term tell the kind of event and the exponents tell the number of that kind of event; e.g., P^2 means there are two P events.

D. Binomial distribution. The binomial distribution may be generated for any N, P and Q by using the binomial expansion. Note that:
1. This distribution is symmetrical with $P = .50$.
2. As N increases, the distribution approximates a normal curve.

E. Binomial table. This table shows the results for the binomial expansion solved for many values of N, P or Q, and no. of P or Q events. Table can be used with P or Q.

III. Using the Normal Approximation.

A. Use. Used instead of the binomial table when $N > 20$. Requires that $NP \geq 10$ and $NQ \geq 10$. When these conditions are met, the binomial distribution approximates the normal curve closely enough to use the normal curve to solve binomial problems.

B. Parameters of Approximate Normal Curve.

$= NP$ *Mean of the approximate normal distribution*

$\sigma = \sqrt{NPQ}$ *Standard deviation of the approximate normal distribution*

C. Steps to Solve Binomial Problems Using the Normal Approximation.

1. If $N \leq 20$, use the binominal table. If $N > 20$, and if the NP and NQ requirements are met, use the normal approximation.
2. Assuming $N > 20$, determine if criteria of $NP \geq 10$ and $NQ \geq 10$ are met.
3. If the criteria in step 2 are met, then compute and of the approximate normal curve.
4. Draw the picture and locate the important information on it.
5. Determine the z value of the outcome specified in the problem.

6. Determine the probability of getting the outcome specified in the problem or any more extreme using the z value and Table A.

CONCEPT REVIEW

The binomial distribution is a (1) _______	(1) probability
distribution. A probability distribution tells us	
each possible (2) _______ and the	(2) outcome
(3) _______ of getting each of the outcomes.	(3) probability
(4) _______ conditions are necessary to	(4) Four
utilize the binomial distribution. They are:	
1. There is a series of (5) _______	(5) N
(6) _______;	(6) trials
2. Each trial has only (7) _______ possible	(7) two
outcomes;	
3. The outcomes are (8) _______	(8) mutually
(9) _______;	(9) exclusive
4. There is (10) _______ between the	(10) independence
outcomes of each trial.	
Tossing a coin is an example of a	

binomial event. If you toss two coins

there are four possible outcomes. They are
(11) _______, (12) _______, (13) _______, | (11) HH
(12) *HT*
(13) *TH*

or (14) _______. The probability | (14) *TT*

of getting two heads is (15) _______. The | (15) 1/4 or *.25*

probability of getting two tails is

(16) _______. From this information | (16) 1/4 or .25

we can construct the probability distribution.

For tossing two coins the following

table results:

Outcome	f	Probability
2 heads	(17) _____	(18) _____
1 head, 1 tail	(19) _____	(20) _____
2 tails	(21) _____	(22) _____

(17) 1
(18) 0.25
(19) 2
(20) 0.50
(21) 1
(22) 0.25

From this table we can determine each of the

possible (23) _______ and the | (23) outcomes

(24) _______ of obtaining | (24) probabilities

those outcomes. For Ns > 2 we can continue

this process of (25) _______ to determine the outcomes and their probabilities. - | (25) enumeration

Another way of obtaining the binomial

distribution for a given size N is

to use the (26) _______ (27) _______. The | (26) binomial (27) expansion

binomial expansion is given by ((28) _______ +. | (28) P

(29) _______)^N. P equals the (30) _______ of | (29) Q (30) probability

any one of the two possible outcomes on a

trial, while (31) _______ equals | (31) Q

the probability of the other possible outcome.

N is the (32) _______ of (33) _______. | (32) number (33) trials

For the example of two coins,

the value $P = Q =$(34) _______. For two tosses | (34) 0.50

of the coin the expansion equals

P^2 +2(35) _______ + Q^2. The terms | (35) PQ

(36) _______, (37) _______, and	(36) P^2 (37) $2PQ$
(38) _______ represent all the possible	(38) Q^2
outcomes of flipping the (39) _______ coins	(39) two
once. The term P^2 tells us this outcome	
consists of 2 P events or 2 (40) _______.	(40) heads
The $2PQ$ term tells us that the outcome	
consists of (41) _______	(41) one
head and (42) _______ tail. The Q^2 tells us the	(42) one
outcome consists of (43) _______	(43) 0
heads. If no heads occurred then the	
outcome must be two (44) _______. If we	(44) tails
wanted to know the probability of getting	
one head and one tail we would solve the	
(45) _______ term. Substituting the value	(45) $2PQ$
of (46) _______ for P and Q, the expression	(46) 0.50
becomes (47) _______((48) _______)	(47) 2 (48) 0.50
((49) _______), which equals (50) _______.	(49) 0.50 (50) 0.50

Fortunately, the results of the binomial expansion are summarized in Table (51) _______. (52) _______ is given in the first column and the possible (53) _______ are given in the second column.

(51) B
(52) *N*
(53) outcomes

The rest of the columns contain probability values for various values of (54) _______.

(54) *P* or *Q*

For example, if one wanted to know what the probability of getting 6 heads in one toss of 10 unbiased coins, one would go to table entry:

N	No. of *P* Events	*P*
(55) _____	(56) _____	(57) _____

(55) 10
(56) 6
(57) 0.50

where one finds the probability equals (58) _______.If one wants to find the probability of obtaining6 or more heads in one toss of the 10 coins,

(58) 0.2051

one simply (59) _______ the probabilities	(59) adds
found in the table for (60) _______,	(60) 6
(61) _______, (62) _______, (63) _______,	(61) 7 (62) 8 (63) 9
and (64) _______ *P* events.	(64) 10
Table B can also be used for values	
of (65) _______ ≠ .50. If $P = 0.30$ and	(65) *P*
$N = 12$, the p(8 *P* events) = (66) _______	(66) 0.0078
If $P > 0.50$, the problem is solved by using	
(67) _______. For example, if $P = 0.85$ and	(67) *Q*
$N = 7$, to find p(5 *P* events), the table	
is entered under:	

N	No. of *Q* Events	*Q*
(68) _____	(69) _____	(70) _____

(68) 7
(69) 2
(70) 0.15

Thus, p(5 *P* events) = (71) _______.	(71) 0.2097
If $N > 20$, the normal approximation may	
be used if $NP \geq$ (72)._______ and *NQ*	(72) 10
(73) _______. In using the normal	(73) 10

approximation, the (74) _______ value of the outcome specified in the problem is	(74) *z*
computed, using = (75) _______ and =	(75) *NP*
(76) _______. The desired probability is	
found by entering Table (77) _______	(76) $\sqrt{NPQ}$
with the *z* value previously computed.	(77) A

EXERCISES

1. What is the probability of tossing 8 unbiased coins once and obtaining at least 5 heads?

2. What is the probability of tossing 6 unbiased coins once and getting 6 heads? What is the probability of getting 6 tails? What is the probability of obtaining either 6 heads or 6 tails?

3. What is the probability of rolling a die and getting an even number exactly 5 times out of 9 rolls?

4. Develop the binomial distribution for the experiment of tossing 3 unbiased coins once. Use the following table to help you set up the problem.

Outcome	**f**	**Probability**
0 heads	_____	_____
1 head	_____	_____
2 heads	_____	_____
3 heads	_____	_____

5. If you were at a race track and bet on 9 races in a day, each with 10 horses entered, what is the probability of winning exactly 2 races if you were picking your winners by guessing alone?

6. If you flipped 8 coins once what is the probability of getting results more extreme than 6 heads? Assume the probability of a head with each coin = 0.75.

7. If you weighted a coin such that the probability of obtaining a head on any one flip was 0.3, what is the probability of getting 5 or 6 heads out of 6 flips? What is the probability of getting 0 or 1 heads out of 6 flips?

8. A coin is said to be biased if $p(H) \neq p(T) \neq 0.50$. If unfair coins are biased such that $p(H) = 0.35$, what is the probability of getting 5 or more heads if 10 coins were tossed once?

9. What is the probability of obtaining 2 or fewer heads if $p(H) = 0.2$ and 7 coins were tossed once?

10. If $p(H) = 0.10$ and 12 coins were flipped once, what are the probabilities for the following outcomes?
    a. Obtaining exactly 3 heads?
    b. Obtaining 0 heads?
    c. Obtaining 4 or fewer heads?
    d. Obtaining results as extreme or more extreme than 4 heads?

11. If $p(H) = 0.75$ what is the probability that if 9 coins were tossed once that one would get 6 or more heads?

12. If $p(H) = 1.00$ and N coins were tossed, how many coins would come up tails?

13. If $p(H) = 0.90$ and 8 coins were tossed once, what is the probability of getting results as extreme or more extreme than 7 heads?

14. If $p(H) = 0.80$ and 9 coins were tossed once, what is the probability of getting results as extreme or more extreme than 3 heads?

15. If the probability of a defective chair = 0.85 and you randomly sample 20 chairs from 10,000 chairs, what is the probability that at least 19 chairs will be defective?

16. A particular industry maintains that from the outset within it women have had the same chance of being hired as men. If this is so, of 10,000 companies within the industry that employ 14 individuals, how many companies would you expect to have 13 or more male employees?

17. If $p(H) = 0.35$ and 30 coins were tossed once, what is the probability of getting 16 or more heads?

18. A manufacturer of ball point pens reports the probability of a defective pen equals 0.20. If you randomly sample 60 pens from 5000 of their pens, what is the probability that 3 or fewer will be defective?
19. You suspect that many of your fellow students are upset over the high cost of gasoline. If the actual percentage is 85% and you randomly sample 100 of your fellow students, what is the probability that at least 90 would be upset over the high cost of gasoline?

TRUE-FALSE QUESTIONS

T F 1. In order to correctly apply the binomial distribution, one of the conditions that must be met is that there are only two possible outcomes.

T F 2. To apply the binomial distribution, three of the conditions which must be met are that there is a series of N trials where the outcomes are mutually exclusive and there is independence between trials.

T F 3. A valid example of a situation where one can apply the binomial distribution is in determining the probability of rolling a 6 or a 5 with the toss of a pair of dice.

T F 4. One can appropriately apply the binomial distribution if $P = 0.37$ and $Q = 0.63$.

T F 5. One can appropriately apply the binomial distribution if $P = 0.5$ and $Q = 0.3$ in a problem.

T F 6. To apply the binomial distribution properly $(P + Q)$ must equal 1.00.

T F 7. $(P + Q)^2 = P^2 + Q^2$ in the binomial expansion.

T F 8. If $P \neq Q$, the binomial distribution will still be normally distributed.

T F 9. The binomial distribution applies equally well to discrete and continuous variables.

T F 10. In $(P + Q)^7$ the term Q^7 indicates 0 P events.

T F 11. The probability of 6 P events when $N = 10$ and $P = 0.7$ is the same as 4 Q events if $Q = 0.3$ and $N = 10$.

T F 12. If $N > 20$, the binomial distribution cannot be used.

T F 13. If $N = 18$ the probability of obtaining a result as extreme or more extreme than 16 P events equals $p(16) + p(17) + p(18)$.

T F 14. For biased coins, the probability of getting 5 heads out of a toss of 5 coins equals the probability of getting 5 tails out of a toss of 5 coins.

T F 15. One can look at picking a winner or not picking a winner in a series of races at the track as fitting the requirements of the binomial distribution (assuming that each horse has an equal chance of winning a particular race and there are the same number of horses in each race).

T F 16. The probability of getting a result as extreme or more extreme than 5 heads out of a toss of 7 unbiased coins is 0.4532.

T F 17. To use the normal approximation, it is required that either $NP \geq 10$ or $NQ \geq 10$, but not both.

T F 18 A smaller N is required to use the normal approximation if $P = 0.60$ than if $P = 0.80$.

SELF-QUIZ

1. Solving thc binomial expansion lo 4 decimal places will give more accurate answers than Table B.

   a. True
   b. False

2. If an event has 3 possible outcomes, then one cannot use the binomial distribution in analyzing the results.
   a. True
   b. False

For problems 3 and 4 consider the binomial expansion for 6 events that is shown below:

$$P^6 + 6P^5Q + 15P^4Q^2 + 20P^3Q^3 + 15P^2Q^4 + 6PQ^5 + Q^6$$

3. What term would you use in evaluating the probability of obtaining exactly 2 heads as the result of flipping an unbiased coin 6 times?

   a. P^6
   b. $20P^3Q^3$
   c. $15P^2Q^4$
   d. the entire expression

4. What term(s) would you use to evaluate the probability of obtaining 1 or fewer heads from 6 flips of an unbiased coin?

   a. $6PQ^5$
   b. $6PQ^5 + Q^6$
   c. Q^6
   d. $20P^3Q^3$

5. If $Q = 0.30$ then P equals _______.

   a. 0.30
   b. 0.90
   c. 0.70
   d. 0.10

6. What is the value of $6P^5Q$ if $P = 0.20$?

   a. 0.0003
   b. 0.1600
   c. 0.0469
   d. 0.0015

7. In order to use the binomial distribution, which of the following conditions are necessary?

   a. a series of N trials with only 2 possible outcomes
   b. outcomes are mutually exclusive
   c. there must be independence between trials
   d. all of the above

8. If $P = 0.40$ for the probability of getting a head on any one flip of the coin, then _______.
   a. the odds of getting 4 out of 4 heads equals the odds of getting 4 out of 4 tails
   b. the odds of getting a head on any one toss is greater than the odds of getting a tail
   c. the odds of getting a tail is greater than the odds of getting a head
   d. cannot be determined

9. In the binomial expansion of $(P + Q)^{10}$, the last term of the expansion will be _______.

   a. Q^{10}
   b. P^{10}
   c. $P^{10} + Q^{10}$
   d. depends on the values of P and Q

10. The sum of all terms in any binomial expansion will equal _______.

    a. 1.0000
    b. 0.5000
    c. depends on the values of P and Q
    d. 0.0000

11. P^0 equals _______.

    a. 1.00
    b. 0.00
    c. 0.50
    d. Q

12. If one flips 5 unbiased coins once, what is the probability of getting exactly 3 heads?

    a. 0.6000
    b. 0.3125
    c. 0.4687
    d. 0.0312

13. What is the probability that one can call the flip of a coin correctly at least 6 out of 7 times assuming that the coin is fair?

    a. 0.0078
    b. 0.0547
    c. 0.5000
    d. 0.0625
    e. 0.1250

14. What is the probability of 4 heads turning up out of four tosses of the coin if the probability of any one head is $P = 0.40$?

    a. 0.0625
    b. 0.1296
    c. 0.0256
    d. 0.3456

15. In one flip of 10 unbiased coins, what is the probability of getting 8 or more heads?

    a. 0.0547
    b. 0.0016
    c. 0.0439
    d. 0.0010

16. In one flip of 10 unbiased coins, what is the probability of getting a result as extreme or more extreme than 8 heads?

    a. 0.0547
    b. 0.1094
    c. 0.0020
    d. 1.0000

17. If one of the terms in a binomial expansion were $210P^6Q^4$, what is the value of N?

    a. 210
    b. 6
    c. 4
    d. 10

18. What is the probability of obtaining exactly 5 P events if the appropriate term of the binomial expansion to evaluate were $6P^5Q^1$ and $P = 0.42$?

    a. 0.0784
    b. 0.0455
    c. 0.4616
    d. 0.0076

19. The probability of obtaining a result as extreme or more extreme than 4 heads if 11 coins were tossed and $p(H) = 0.35$ is _______.

    a. 0.6683
    b. 0.2745
    c. 0.5490
    d. 0.7184

20. A football player is practicing making field goals from the 30-yard line. If the probability of his kicking a field goal is 0.75, what is the probability he will kick at least 12 field goals in the next 15 tries? Assume independence between tries.

    a. 0.0000
    b. 0.2252
    c. 0.4613
    d. 0.6481

21. In solving a binomial problem, if the evaluation required only finding the probability of all P events, and $N = 80$, evaluating P^{80} would be more accurate than using the normal approximation. Assume no rounding.

    a. T
    b. F

22. Evaluated over a large number of pitches, a varsity women softball pitcher threw 1400 strikes and 600 balls. Assuming things remain the same, what is the probability she will throw at least 35 strikes in her next 40 pitches. Assume independence between throws.

    a. 0.0427
    b. 0.0418
    c. 0.0001
    d. 0.0080

ANSWERS

Exercises: 1. 0.3633 **2.** 0.0156, 0.0156, 0.0312 **3.** 0.2461

4. binomial distribution table:

f	Probability
1	0.1250
3	0.3750
3	0.3750
1	0.1250

5. 0.1722 **6.** 0.3671 **7.** 0.0109, 0.4201 **8.** 0.2485 **9.** 0.8520 **10a.** 0.0852 **10b.** 0.2824 **10c.** 0.9956 **10d.** 0.9956 **11.** 0.8343 **12.** 0 **13.** 0.8131 **14.** 0.9175 **15.** 0.1756 **16.** 10. **17.** $z = 2.11$; $p = 0.0174$ **18.** $z = 2.90$; $p = 0.0019$ **19.** $z = 1.40$; $p = 0.0808$.

True-False: **1.** T **2.** T **3.** F **4.** T **5.** F **6.** T **7.** F **8.** F **9.** F **10.** T **11.** T **12.** F **13.** F **14.** F **15.** T **16.** T **17.** F **18.** T.

Self-Quiz: **1.** b **2.** a **3.** c **4.** b **5.** c **6.** d **7.** d **8.** c **9.** a **10.** a **11.** a **12.** b **13.** d **14.** c **15.** a **16.** b **17.** d **18.** b **19.** a **20.** c. **21.** a. **22.** $z = 2.42$; $p = 0.0078$.

Chapter 10

Introduction to Hypothesis Testing Using the Sign Test

CHAPTER OUTLINE

I. Logic of Hypothesis Testing

A. The experiment.

1. Purpose of an experiment. An experiment allows the scientist to test the influence of the independent variable upon the dependent variable while controlling for the influence of other variables.
2. Conclusions. In making a conclusion based upon an experiment, one has to answer the question, "How reasonable are these results if chance alone were responsible for the results?" If the results are not likely to be due to chance, then the results are attributed to the experimental manipulation.

B. Repeated measures design. This is one of several commonly used designs. It is also called the replicated measures design or correlated groups design. The essential feature is that subjects are paired prior to conducting the experiment and the difference between paired scores are analyzed.

C. Alternative hypothesis (H_1). The alternative hypothesis claims that the difference in results between conditions is due to the independent variable.

1. Directional hypothesis. The hypothesis specifies the direction of the effect of the independent variable; e.g., marijuana increases appetite.
2. Nondirectional. The hypothesis states that the independent variable has an effect but the direction of the effect is not stated.

D. Null hypothesis (H_0). The null hypothesis is the logical counterpart of the alternative hypothesis. For a nondirectional alternative hypothesis, the null hypothesis states that the independent variable has no effect on the dependent variable. For a directional alternative hypothesis, the null hypothesis states that the independent variable has no effect in the direction predicted by H_1.

E. Mutually exclusive and exhaustive. The alternative hypothesis and the null hypothesis are mutually exclusive and exhaustive. If one is true the other cannot be. If one is false the other must be true. We always analyze the null hypothesis and try to show that it is false. If we show the null hypothesis is false, then we can accept the alternative hypothesis as true.

F. Decision rule. We evaluate H_0 directly because we can calculate the probability of chance events, but there are no mathematics for the probability of the alternative hypothesis.

1. Chance. We calculate the probability of the obtained results if chance alone were operating.
2. Alpha level (α). If the resulting probability turns out to be less than or equal to a critical probability level called the alpha level, we reject the null hypothesis. Common levels for the alpha are $\alpha = 0.05$ and $\alpha = 0.01$. When we reject H_0, the results are said to be significant or reliable.

II. Possible Decision Errors

A. Type I error. This is when the null hypothesis is rejected when in fact the null hypothesis is true.

B. Type II error. This is when one retains the null hypothesis when in fact the null hypothesis is false.

C. Trade-offs. The probability of making a Type I error is set by alpha. The probability of making a Type II error is called beta. If we make alpha more stringent, we increase beta. Setting alpha and beta depends upon what it would cost to make either a Type I or Type II error.

III. Evaluating the Tail of the Distribution

A. Directional hypothesis. If the alternative hypothesis is directional, we determine the probability of getting the obtained outcome or any even more extreme in the direction hypothesized. We evaluate the tail of the distribution.

B. Nondirectional hypothesis. If H_1 is nondirectional, we evaluate the obtained result or any even more extreme in both directions (both tails).

C. Two-tailed probability evaluations. If the experimenter has no valid basis for directional hypothesis, one uses a two-tailed evaluation.

D. One-tailed probability evaluation. If the experimenter has a good theoretical basis and data from other studies, a directional hypothesis may be used and the experimenter may use a one-tailed probability evaluation.

IV. Summary of Hypothesis Testing Using the Sign Test

A. Process.

1. Used only in replicated measures design.
2. Assumptions of binomial distribution must be satisfied.
3. Null and alternative hypotheses generated. H_1 is directional or nondirectional,
4. Alpha level set.
5. Difference between control and experimental conditions is calculated.
   a. Sign of the difference is recorded.
   b. Magnitude of difference is ignored
   c. Ties are ignored.
6. Calculate the probability of getting the obtained results and results more extreme using one or two tails of the binomial distribution depending on the nature of alternative hypothesis and alpha level.
7. Compare resulting probability to alpha.
8. Draw conclusions and generalizations (with the appropriate cautions).

V. Size of Effect

A. Concepts.

1. The size of effect refers to how large or small the real effect of the independent variable is on the dependent variable.

2 Statistically significant results are statistically reliable. Because an effect is statistically significant, it doesn't necessarily follow that the effect is an important one. Important effects often depend on the size of the effect as well as its reliability.

CONCEPT REVIEW

When a scientist has an idea, he can generate a (1) _______ which can be tested in an (2) _______. An experiment allows the scientist to test the influence of the (3) _______ variable upon the (4) _______ variable while controlling for the (5) _______ of other variables.

(1) hypothesis
(2) experiment
(3) independent
(4) dependent
(5) influence

When a scientist asks an experimental question it generally centers around an (6) _______ he has developed. The (7) _______ hypothesis states that the independent variable (8) _______ the dependent variable. In every experiment, besides the alternative hypothesis, there exists

(6) hypothesis
(7) alternative
(8) affects

the (9) _______ hypothesis. The null hypothesis for a nondirectional alternative hypothesis asserts that the independent variable has (10) _______ effect upon the dependent variable and the observed results are due to (11) _______. These two hypotheses are logical counterparts to each other. They are (12) _______ (13) _______ and (14) _______. If one hypothesis is true the other must be(15) _______. If one is false the other must be (16) _______. The alternative hypothesis is symbolized by (17) _______. The null hypothesis is symbolized by (18) _______. The alternative hypothesis can be either (19) _______ or (20) _______. In a directional hypothesis the experimenter specifies the (21) _______ of the effect of the independent variable upon the dependent variable.

(9) null
(10) no
(11) chance
(12) mutually
(13) exclusive
(14) exhaustive
(15) false
(16) true
(17) H_1
(18) H_0
(19) directional
(20) nondirectional
(21) direction

In a nondirectional hypothesis the experimenter merely states that there is an (22) _______, but not in which (23) _______.

(22) effect
(23) direction

After the experimenter conducts an experiment, he evaluates the (24) _______ hypothesis by determining the (25) _______ of obtaining the observed results or results even more extreme if chance alone is at work. The reason we evaluate the null hypothesis is because we can calculate the probability of (26) _______ events but there are no mathematics worked out for the (27) _______ hypothesis. However, if we can show that the null hypothesis is (28) _______, then the alternative must be true.

(24) null
(25) probability
(26) chance
(27) alternative
(28) false

To reject the null hypothesis, one must make a decision depending on the likelihood of the observed results (tail evaluation). If the probability of observing the outcome of the study is too low to be reasonably explained by chance, we can reject H_0.

By convention, if the probability of observing the results by chance is equal to or less than (29) _______ or (30) _______, we reject (31) _______. The value 0.05 or 0.01 is called the (32) _______ (33) _______. If the probability of observing the obtained results is (34) _______ or (35) _______ the (36) _______level, we reject the (37) _______ hypothesis. If the alpha level equals 0.05, then the null hypothesis would be rejected if the observed results occur with a probability equal to or less than (38) _______ times out of (39) _______ assuming (40) _______ alone.

(29) 0.05
(30) 0.01
(31) H_0
(32) alpha
(33) level
(34) equal to
(35) less than
(36) alpha
(37) null
(38) five
(39) 100
(40) chance

When one evaluates the outcome of an experiment one tries to determine the state of reality concerning the independent variable.

An experiment allows us to make an inference about that reality but our inferences could be wrong. If we reject H_0 when H_0 is false we have made a (41) _______ decision. If we retain H_0 when H_0 is true, we have also made a (42) _______ decision. If we reject H_0 when H_0 is true, we have made a(43) _______ error. If we retain H_0 when H_0 is false, we have made a (44) _______ error. The probability of a Type I error is determined by (45) _______. The probability of a type II error is given by (46) _______. Unfortunately, changing alpha affects (47) _______. As alpha is made more stringent, beta (48) _______. Thus, decreasing the probability of a Type I error increases the probability of a Type (49) _______ error. (Life is like that sometimes.)

(41) correct
(42) correct
(43) Type I
(44) Type II
(45) alpha
(46) beta
(47) beta
(48) increases
(49) II

In analyzing the results of a particular study,

it is (50) _______ to analyze just the specific	(50) incorrect
outcome. Instead, we must determine the	
(51) _______ of getting the specific outcome	(51) probability
or any even more (52) _______. It is this	(52) extreme
probability we compare to(53) _______ to	(53) alpha
assess the reasonableness of the (54) _______	(54) null
(55) _______. This is to say that we evaluate	(55) hypothesis
the (56) _______ of the distribution. If H_1 is	(56) tail
nondirectional we evaluate (57) _______ tails.	(57) both
As an example of a statistical inference test	
we have introduced the (58) _______ test.	(58) sign
It is used in (59) _______ (60) _______	(59) repeated (60) measures
designs. Data are generally collected under	
a (61) _______ and (62) _______ condition	(61) control (62) experimental
most often on the same (63) _______. The	(63) subjects
(64) _______ between these conditions is	(64) difference
calculated and the number of (65) _______	(65) pluses
and (66) _______ are recorded.	(66) minuses

The (67) _______ of the differences is	(67) magnitude
ignored. The sign test is not very (68) _______	(68) sensitive
because it does ignore the magnitude of the	
differences. In the case of ties, the ties are	
(69) _______. One then evaluates the number	(69) discarded
of pluses (or minuses) just like heads (or tails)	
using the (70) _______ distribution. This	(70) binomial
presupposes all of the (71) _______ of the	(71) assumptions
binomial distribution. One then calculates	
the probability of getting the obtained results	
or results (72) _______(73) _______.	(72) more (73) extreme
Whether one evaluates one or both	
(74) _______ of the distribution depends on	(74) tails
whether the alternative hypothesis is	
(75) _______ or (76) _______. One then	(75) directional (76) nondirectional
compares the resulting probability with	
(77) _______. If the obtained probability is	(77) alpha
less than (78) _______ we (79) _______	(78) alpha (79) reject

(80) _______. If one rejects the null hypothesis, one is at risk of making a	(80) H_0
(81) _______ error. If one retains the null hypothesis one may be making a	(81) Type I
(82) _______ error. Notice that we always	(82) Type II
reject or retain the (83) _______ hypothesis. Finally, one can draw conclusions and offer generalizations. One can only generalize to the	(83) null
(84) _______ from which the	(84) population
(85) _______ was drawn.	(85) sample
In an experiment there are only two possibilities. Either H_0 is (86) _______ or	(86) true
it is (87) _______. By minimizing	(87) false
(88) _______ and (89) _______ we maximize the likelihood our conclusions will be	(88) alpha (89) beta
(90) _______.	(90) correct
A statistically significant result means the result is (91) _______. Before a result can be	(91) reliable

considered important, it must be reliable.

However, (92) ______ effects as well as large effects can be statistically significant, and often small effects are not (93) _______ ones. Therefore, a statistically (94) _______ result is not necessarily an (95) _______ one. Often the (96) _______ of the effect must be considered too.

(92) small
(93) important
(94) significant
(95) important
(96) size

EXERCISES

1. A teacher wanted to compare two methods for teaching math. He randomly divided the class in half. Half the class was given the standard method first, followed by the new method. The other half was given the two methods in the opposite order. The results of tests given after each of the two methods are shown in the following table:

Student	**Standard Method**	**New Method**
1	**80**	**82**
2	**78**	**79**
3	**65**	**75**
4	**92**	**100**
5	**85**	**85**
6	**60**	**61**
7	**64**	**60**
8	**82**	**80**
9	**90**	**92**
10	**78**	**86**
11	**70**	**79**
12	**85**	**90**
13	**86**	**80**

a. State the nondirectional alternative hypothesis.
b. State the null hypothesis.
c. Using □ = 0.052 tail, what do you conclude?
d. What type of error might you be making?

2. A panel of consumers was asked to rate "Natural Vitaglo" breakfast cereal before and after a change in manufacturing techniques. A higher number reflects a higher preference for the cereal. The results are given in the following table.

Rater	Old Process	New Process
A	6	10
B	7	9
C	7	8
D	8	10
E	5	6
F	9	10
G	10	9
H	4	5
I	6	8
J	3	4

a. State the nondirectional alternative hypothesis.
b. State the null hypothesis.
c. What do you conclude using □ = 0.052 tail?
d. What type of error might you be making?
e. What would your conclusion be if □ = 0.012 tail had been specified?
f. Why did the investigator use a 2-tailed test?

3. Why do we directly test the null hypothesis instead of the alternative hypothesis?

4. A young man in a pinstriped suit, black shirt and white tie comes up to you and asks if you want to flip a coin for a

dollar. Feeling lucky you agree. You call "tails" eight times in a row and you lose each time. You begin to wonder if the coin is biased.

a. State the directional alternative hypothesis.
b. State the null hypothesis.
c. What do you conclude using □ = 0.051 tail?
d. Why did you start playing with that guy anyway?

5. If you were a scientist who believes she has discovered an inexpensive new energy source that doesn't pollute, and you designed an experiment to assess whether or not this energy source actually worked, would you rather protect yourself from making a Type I or Type II error? Why?

6. A social scientist wants to determine if showing a film will affect religious tolerance. A sample of freshman males was drawn from a university. Pre and post film ratings of religious tolerance were obtained. A higher number represents higher tolerance. The following data were obtained:

Person	Pre Film	Post Film
1	50	89
2	52	92
3	61	93
4	70	68
5	58	95

a. State the nondirectional alternative hypothesis.
b. State the null hypothesis.
c. What do you conclude using □ = 0.012 tail?
d. To what population can you generalize?
e. Are there any results from a sample of the size that would have allowed you to reject H0?

7. A pharmacologist recently developed a drug that, in theory, should have an effect on the activity of the lateral hypothalamus. It is likely that such a drug would have an effect on appetite. A small pilot experiment is designed to test this hypothesis.
   a. State H0 and H1.
   b. At what level would you set alpha? Why?
   c. Would you use a one- or two-tailed test? Why?
8. An after-shave company recently advertised that if you use their product you will have more sex appeal. To test this claim you enlist the assistance of 12 men from a local social club. The men are either given the new after-shave or colored water. After one month the type of liquid each man had been using was changed to the other type; i.e., water changed for after-shave or after-shave for water. The experiment continued for another month. The dependent variable was the number of dates each man had for each month. The results are shown in the following table

Subject	**1**	**2**	**3**	**4**	**5**	**6**	**7**	**8**	**9**	**10**	**11**	**12**
Colored water	**7**	**9**	**4**	**2**	**8**	**3**	**6**	**5**	**10**	**1**	**0**	**4**
After-shave	**6**	**5**	**10**	**3**	**9**	**7**	**5**	**5**	**11**	**2**	**1**	**5**

   a. State H0 and H1.
   b. What do you conclude from the data shown above? Use □ = 0.052 tail
   c. What type of error might you have made?
   d. What is one possible explanation for these results?

9. The following probabilities are the results of several different experiments.

Experiment	$p(\text{obtained})_{2\text{ tail}}$
1	.01
2	.04
3	.05
4	.17
5	.20
6	.006

a. What would you conclude for each experiment if $\alpha = 0.05_{2\text{ tail}}$?
b. What would you conclude if $\alpha = 0.01_{2\text{ tail}}$?

10. What are the possible costs to making a Type I error in a study that tests the null hypothesis that a new type of male contraceptive pill is not effective? (The alternative hypothesis is that it is effective.)

TRUE-FALSE QUESTIONS

T F 1. The sign test ignores the magnitude of the difference scores.

T F 2. The sign test analyzes raw scores.

T F 3. It is impossible to get 20 pluses out of 20 pairs of scores in a replicated measures design due to chance alone.

T F 4. A replicated measures design is the same thing as a correlated groups design.

T F 5. Unless the very same subject is used in the control and experimental condition the design cannot be a replicated measures design.

T F 6. The statement "Drug *X* has no effect on *Y* and any observed effect is due to chance alone" is an example of a nondirectional alternative hypothesis.

T F 7. H_0 and H_1 must be mutually exclusive and exhaustive.

T F 8. H_0 always asserts that the dependent variable has no effect on the independent variable.

T F 9. It is not possible to analyze the probability of the alternative hypothesis because the laws of probability are derived for chance events. This is why we test H_0 instead of H_1.

T F 10. If we reject H_0 one can say the experimental results are significant.

T F 11. If we reject H_0 then we are absolutely certain the independent variable affected the dependent variable.

T F 12. The critical probability for rejecting H_0 is the alpha level.

T F 13. Retaining H_0 and accepting H_0 mean the same thing.

T F 14. Type I errors are always worse to make than Type II errors.

T F 15. If H_0 is true and you retain H_0 in your experiment you have made a Type II error.

T F 16. If the probability of the results obtained equals 0.05 and $\alpha = 0.05$, you should reject H_0.

T F 17. By making alpha smaller we can decrease the probability of making a Type I error.

T F 18. Beta is the probability of retaining H_0 when H_0 is false.

T F 19. The more stringent the alpha level is, the easier it is to detect an effect of the independent variable on the dependent variable.

T F 20. If $\alpha = 0.05$ and chance alone is operating, it is reasonable to expect that on the average one would make one Type I error in every 20 experiments.

T F 21. As the probability of making a Type I error goes down by making α more stringent, the probability of making a Type II error goes up.

T F 22. In the real world the scientist can never be certain whether or not H_0 is true.

T F 23. The probability of making a Type I error increases with independent replication.

T F 24. In hypothesis testing it is incorrect to evaluate the probability of the specific outcome of the experiment.

T F 25. If H_1 is nondirectional one must evaluate the probability of getting the obtained result or a result more extreme in both directions to properly test H_0.

T F 26. It is appropriate to determine if H_1 is directional or nondirectional after the data is analyzed.

T F 27. If $\alpha = 0.01$ and H_1 is 2-tailed then there is 0.005 under each tail.

T F 28. Beta is the probability of retaining H_0 when H_0 is false.

T F 29. If a result is statistically significant, it must be an important result.

SELF-QUIZ

1. If p (obtained) from an experiment equals 0.05 and alpha equals 0.05 (both two-tailed), what would you conclude?

   a. reject H0
   b. retain H0
   c. reject H1
   d. retain H1

2. If you reject the null hypothesis, what type of error might you be making?

   a. Type I
   b. Type II
   c. Type III
   d. cannot be determined

3. If alpha equals 0.05, how many times out of 100 would you expect to reject the null hypothesis when the null hypothesis is in fact true?

   a. 1
   b. 0.05
   c. 0.01
   d. 5

4. If we drew a random sample from an introductory psychology class, to whom could we generalize our results?

a. all of human kind
b. the university
c. all psychology students
d. the students in that introductory psychology class

5. If we set alpha at 0.05 instead of 0.01, other factors held constant _______.

   a. we have a greater risk of a Type I error
   b. we have a greater risk of a Type II error
   c. we have a lower risk of a Type II error
   d. a and c

6. If you reject H0 when H0 is false, you have made a _______.

   a. Type I error
   b. Type II error
   c. correct decision
   d. none of the above

7. If the alpha level is changed from 0.05 to 0.01, what effect does this have on beta?

   a. beta decreases
   b. beta increases
   c. beta is unaffected
   d. cannot be determined

8. The sign test can be used for _______.

   a. a repeated measures design
   b. a replicated measures design
   c. a correlated measures design
   d. all of the above

9. One can say it is always preferable to make a Type II error.

   a. True
   b. False
   c. It depends on the costs of making a Type I or Type II error

10. In a nondirectional alternative hypothesis, evaluating the probability of observing 7 pluses out of 8 events equals ________.

    a. p(7)
    b. p(7) + p(8)
    c. p(0) + p(1) + p(7) + p(8)
    d. p(0) + p(8)

Consider the following hypothetical data collected using replicated measures design:

Subject	**1**	**2**	**3**	**4**	**5**	**6**	**7**	**8**	**9**	**10**
Pre	**50**	**49**	**37**	**16**	**80**	**42**	**40**	**58**	**31**	**21**
Post	**56**	**50**	**30**	**25**	**90**	**44**	**60**	**71**	**32**	**22**

11. In a two-tailed test of H0 using □ = 0.05, what is p(obtained) for the results shown?

    a. 0.0500
    b. 0.0108
    c. 0.1094
    d. 0.0216

12. What would you conclude using the information in problem 11?

    a. reject H0
    b. accept H0

c. retain H0
d. retain H1
e. b or c

13. What would you conclude using □ = 0.012 tail?

a. reject H0
b. accept H0
c. retain H0
d. fail to reject H1
e. b or c

14. What type error might you be making using □ = 0.052 tail?

a. Type I
b. Type II
c. Type III
d. cannot be determined

15. If the results of an experiment are statistically significant,

a. the results must important
b. the results should be ignored
c. the results might important
d. the results are reliable
e. a and d
f. c and d

ANSWERS

Exercises: 1a. The new teaching method affects test scores. **1b.** The new teaching method has no effect on test scores. $P = Q = 0.50$. **1c.** Retain H_0 since .1458 is greater than 0.05. This experiment fails to establish that there is any difference in effectiveness between the new and standard teaching methods.

1d. Type II error. (NOTE: did you remember to omit case 5 because of tied scores?)

2a. The new manufacturing process affects the taste of "Natural Vitaglo" as assessed by preference ratings. **2b.** The new manufacturing process has no effect on the taste of "Natural Vitaglo" as assessed by preference ratings. $P = Q = 0.50$. **2c.** Since $0.0216 < 0.05$, reject H_0 in favor of H_1. The change in process affects taste. **2d.** Type I error. **2e.** Retain H_0. **2f.** Nondirectional H_1.
3. Because the mathematics have only been established for chance events.

4a. The coin is biased in favor of producing heads. **4b.** The coin is not biased in favor of producing heads. $P = Q = 0.50$. **4c.** Reject H_0 since 0.0039 is less than 0.05 **4d.** Who knows?
5. There is no correct answer but probably you would wish to protect yourself against a Type II error since if H_0 is false you definitely want to reject it and it seems there are more costs to making a Type II error than a Type I error. On the other hand, to avoid misleading the public, you would probably want to set alpha to at least 0.05.

6a. The film affects religious tolerance. **6b.** The film has no effect on religious tolerance and the results are due to chance alone. **6c.** Retain H_0 since $0.3718 > 0.01$. The experiment fails to show that the film affects religious tolerance. **6d.** freshmen men at that university **6e.** No, it's a very poor experiment.

7a. H_0: the new drug has no effect on food intake. Any effect is due to chance alone. H_1: taking this new drug effects food intake. The differences are such that they cannot reasonably be attributed to chance. **7b.** $\alpha = 0.10$. During a pilot study one generally wishes to set alpha at a less stringent level than perhaps under other circumstances to see if a larger scale trial is warranted. There is no correct answer to this question. You should have a reasonable rationale for whatever alpha level you choose. **7c.** Two-tailed. From the information given there is no justification for selecting a directional hypothesis.

8a. H_1: using the after-shave has an effect on the number of dates men who wear it obtain. H_0: the use of the after-shave has no effect on the number of dates men who wear it obtain. $P = Q = 0.50$. **8b.** $p = 0.2268$ which is $> \alpha$; therefore retain H_0. **8c.** Type II **8d.** The number of dates might not be a good measure of sex appeal. There are a lot of other possible answers.

9a. Reject H_0 for 1, 2, 3, and 6; retain H_0 for 4 and 5. **9b.** Reject H_0 for 1 and 6; retain for 2, 3, 4, and 5.

10. Increase in birth rate, possible increase in marriage rate, unplanned pregnancies, etc., etc.

True-False: **1.** T **2.** F **3.** F **4.** T **5.** F **6.** F **7.** T **8.** F **9.** T **10.** T **11.** F **12.** T **13.** F **14.** F **15.** F **16.** T **17.** T **18.** T **19.** F **20.** T **21.** T **22.** T **23.** F **24.** T **25.** T **26.** F **27.** T **28.** T **29.** F

Self-Quiz: **1.** a **2.** a **3.** d **4.** d **5.** d **6.** c **7.** b **8.** d **9.** c **10.** c **11.** d **12.** a **13.** c **14.** a **15.** f.

Chapter 11

POWER

CHAPTER OUTLINE

I. Definitions

A. Definitions.

1 Power is the probability that the results of experiment will allow rejection of the null hypothesis if the independent variable has a real effect.
2. Power is the probability that the results of experiment will allow rejection of the null hypothesis if the null hypothesis is false.
3. Power is the probability of making a correct decision when H_0 is false.

B. P_{real}.

1. Definition. P_{real} is the probability of a plus with any subject in the sample of the experiment when the independent variable has a real effect. It is also the proportion of pluses in the population if the experiment was done on the entire population and the independent variable has a real effect.
2. P_{real} varies with the size and direction of the real effect.

3. P_{real} equals any value other than 0.50 (H_1 nondirectional)

C. P_{null}.

1. Definition. P_{null} is the probability of getting a plus with any subject in the sample of the experiment when the independent variable has no effect
2. P_{null} = 0.50 (H_1 nondirectional).

II. Power and Beta (β)

A. <u>Formula</u>.

Power = 1 − beta.

Beta = 1 − power.

B. <u>Relationship between power and beta</u>.

1. As the power of an experiment increases, the probability of making a Type II error decreases.
2. Maximizing power minimizes beta.

C. <u>Methods of increasing power</u>.

1. Increase the size of effect of independent variable.
2. Increase sample size.

III. Identifying Correct State of Reality

A. <u>Alpha and beta values</u>.

1. Set alpha stringently (minimize Type I error).

2. Set beta low (minimize Type II error).

IV. Calculation of Power is a Two-Step Process

A. Step 1. Assume the null hypothesis is true (P_{null} = 0.50) and determine the possible sample outcomes in the experiment that allow H_0 to be rejected.

B. Step 2. For the value of P_{real} under consideration (e.g. P_{real} = 0.40) determine the probability of getting any one of the above sample outcomes. This probability is the power of the experiment to detect an effect size equal to P_{real}.

CONCEPT REVIEW

(1) _______ is the probability of making a	(1) Beta
Type II error. By maximizing (2) _______ we	(2) power
can minimize beta. The power of an experiment	
is a measure of the(3) _______ of the experiment	(3) sensitivity
to detect the real (4) _______ of the (5) _______	(4) effect (5) independent
variable. A more formal definition of power is the	
(6) _______ of rejecting (7) _______ when the	(6) probability (7) H_0

(8) _______ variable has a (9) _______. (8) independent (9) real

effect. P_{real} is the probability of getting a

plus if the independentvariable has a

(10) _______ effect. P_{null} is the probability (10) real

of getting a plus if the independent variable

has (11) _______ effect. P_{real} is a measure of (11) no

the (12) _______ and (13) _______ of (12) size (13) direction

the real effect of the (14) _______ variable. (14) Independent

The further (15) _______ is from 0.50, (15) Preal

the (16) _______ is (16) larger

the effect of the independent variable.

The closer the value of (17) _______ (17) Preal

is to 0.50, the (18) _______ is the (18) smaller

(19) _______ effect of the independent variable. (19) real

If $P_{real} = 0.30$, the effect is(20) _______ (20) smaller

than for Preal = 0.20. In a given experiment, the

greater the (21) _______ of the independent variable the greater the (22) _______.

(21) effect
(22) power

To calculate power we first assume that H_0 is (23) _______ (P_{null} = (24) _______). Then determine all the (25) _______ outcomes in the experiment which allow H_0 to be (26) _______ Second, for the magnitude of real effect under consideration, we determine the (27) _______ of getting any one of the above sample outcomes. This probability is the (28) _______ of the experiment to detect this level of (29) _______ effect. For each size of real effect, we must calculate a new value for (30) _______. If the power of an experiment is 0.6500, that means that we have a (31) _______ % chance of rejecting H_0 when H_0 is false and a (32) _______%

(23) true
(24) 0.50
(25) sample
(26) rejected
(27) probability
(28) power
(29) real
(30) power
(31) 65.00
(32) 35.00

chance of making a Type II error. If H_0 is true, the probability of rejecting the null hypothesis is determined by the (33) _______ level.

(33) alpha

In an experiment there are only two possibilities. Either H_0 is (34) _______ or it is (35) _______. By minimizing (36) _______ and (37) _______ we maximize the likelihood our conclusions will be (38) _______. Power = 1− (39) _______.

(34) true
(35) false
(36) alpha
(37) beta
(38) correct
(39) beta

One way to achieve a low beta when alpha is set at a stringent level is to have a larger (40) _______.

(40) *N* (sample size)

If one obtains nonsignificant results in an experiment, one (41) _______ to reject H_0. This may be because H_0 is (42) _______ or because H_0 is false, but the (43) _______

(41) fails
(42) true
(43) power

of the experiment was (44) _______. | (44) low

Therefore, we (45) _______.accept H_0 | (45) do not

Rather, we (46) _______ it as a reasonable | (46) retain

possible explanation of the results. A powerful

experiment will yield a higher likelihood of

rejecting H_0 only if H_0 is (47) _______. | (47) false

A power analysis is useful when initially

(48) _______ an experiment or when | (48) designing

interpreting the (49) _______ results from | (49) nonsignificant

an experiment. Since we never know the

size of the effect of the independent variable

before carrying out the experiment, we usually

(50) _______ a value for P_{real} based on | (50) estimate

pilot work or similar studies.

EXERCISES

1. A professor thinks that a certain movie is fun and exciting for students. He selects heart rate (HR) measured before and after the movie as the dependent variable for a study on 12 students. He figures if the students are having fun and are excited, the heart rate will be higher right after the movie. The following data have been obtained.

Student	1	2	3	4	5	6	7	8	9	10	11	12
Pre-movie HR	70	68	74	88	75	79	95	65	87	73	69	96
Post-movie HR	72	69	71	92	74	85	100	80	90	77	71	88

   a. Based on these results, what should the professor conclude using $\alpha = 0.05_{2\text{ tail}}$? (Assume that heart rate is a good measure of fun and excitement.).
   b. If the effect of the independent variable were such that $P_{\text{real}} = 0.60$, what is the power in this experiment?
   c. What type error might the investigator be making?

2. In an experiment where the experimental effect is such that $P_{\text{real}} = 0.70$ and $N = 14$, what is the power of the experiment? Use $\alpha = 0.05_{2\text{ tail}}$ with the sign test.

3 Assume you have done an experiment and used the sign test to analyze the results. $N = 14$.

   a. If your experimental treatment was moderately effective ($P_{\text{real}} = 0.70$), what was the power of your experiment? Assume $\alpha = 0.05_{1\text{ tail}}$.
   b. Generalize from questions 2 and 3a about alpha and power.

4 Assume you have $N = 15$, $\alpha = 0.05_{1\ \text{tail}}$, and applied the sign test to the data.

a. What is the power if $P_{\text{real}} = 0.80$?
b. If you increased the sample size to 20, what would be the power?
c. What is the value of beta if $N = 20$?
d. Generalize from questions 4a and 4b about N and power.

5. You are considering testing a new drug that is supposed to facilitate learning in mentally retarded children. Based on preliminary research, you have some idea about the size of the drug's effect. Because of your work schedule you can either do the test with 15 subjects or with 20. You would like to only run 15 subjects, but if running 20 will make power at least 20% higher, you will run 20 subjects. Calculate power for the following conditions:

a. $N = 15$, $\alpha = .05_{1\ \text{tail}}$, $P_{\text{real}} = 0.70$.
b. $N = 20$, $\alpha = .05_{1\ \text{tail}}$, $P_{\text{real}} = 0.70$.
c. Based on your calculations, how many subjects will you test?

6. A wine company has retained you as a consultant to help evaluate the taste preference for their wine compared to the market leader. You plan to give 20 subjects a taste of both wines and ask them to rate the wines on a 10 point scale. The order of presentation is randomized so there is no order effect. Your plan is to analyze the results using the sign test. If the size of the difference is such that $P_{\text{real}} = 0.80$, what is the power of your experiment to detect a preference for one wine or the other? Use $\alpha = 0.05_{2\ \text{tail}}$.

7. Ajax Pest Control Company has bred a new strain of pest-eating insects for the garden. In several test gardens the

number of pests before and after introduction of their new insect has been recorded. The following data were obtained:

Garden	1	2	3	4	5	6	7	8
Pre	32	67	83	22	39	44	90	11
Post	26	12	34	66	10	0	17	25

a. What is the power of this experiment using $\alpha = 0.01_{2\ \text{tail}}$ if $P_{\text{real}} = 0.70$?
b. How could you increase the power?

8. In question 7, what is the probability of making a Type II error?

9. In an experiment, if the analysis of the data does not allow us to reject the null hypothesis, this can only be so because we do not have enough experimental power to reject H_0. True or false? Why?

10. In an experiment the calculated power to detect an effect is .4600. What is the value for beta?

11. In an experiment the value for beta has been shown to be .7503. What is the power of this experiment?

TRUE-FALSE QUESTIONS

T F 1. If H_0 is true, power is the probability of not making a Type II error.

T F 2. If in reality H_0 is true, then designing a very powerful experiment (without changing alpha) will allow us to reject H_0 more readily.

T F 3. Power is the probability of making a correct decision when H_0 is true.

T F 4. If the independent variable is more powerful in experiment A than in experiment B, there is a higher probability of rejecting H_0 in A than in B if, in fact, H_0 is false.

T F 5. H_0 states that the independent variable has no effect. H_1 is nondirectional.

T F 6. It is impossible for an experiment to have power equal to zero.

T F 7. An experiment where $P_{real} = 0.10$ is less likely to allow rejection of H_0 if H_0 is false than one where $P_{real} = 0.80$.

T F 8. The greater the sample size the greater the power.

T F 9. Power $= 1 - \beta$

T F 10. Beta is the probability of retaining H_0 when H_0 is false.

T F 11. $\beta = 1 - \alpha$.

T F 12. A good experiment is one where both α and β are low.

T F 13. If one conducts an experiment and ends up retaining H_0, it is probably because the experiment was not powerful enough.

T F 14. In an experiment where we retain H_0, one cannot be certain if it was because H_0 was true or that the experiment was not powerful enough to detect an effect by the independent variable.

T F 15. The concept of statistical power applies only to the sign test.

T F 16. An experiment with $\alpha = 0.05$ is more powerful than one where $\alpha = 0.01$, other factors held constant.

T F 17. Increasing N, making α more stringent, and increasing the effect of the independent variable, all increase power.

SELF-QUIZ

1. If the independent variable has a real effect, the probability of rejecting H_0 is _______.

   a. power
   b. 1 − power
   c. alpha
   d. beta

2. One can increase power by _______.

   a. increasing N
   b. making alpha more stringent
   c. increasing the effect of independent variable
   d. a and c

3. If the power of an experiment is 0.3400, the probability of retaining H_0 when H_0 is false is _______.

   a. 0.6600
   b. 0.3400
   c. 0.5000
   d. 1.0000

4. Which of the following *P* values represents the strongest effect?

   a. $P_{null} = 0.50$
   b. $P_{real} = 0.70$
   c. $P_{real} = 0.80$
   d. $P_{real} = 0.10$

5. If beta = 0.7500, what is the power of the experiment?

   a. 0.7500
   b. 0.5000
   c. 0.2500
   d. 1.0000

6. Which of the following represents the null hypothesis condition for a nondirectional H_1?

   a. $P_{real} = 0.49$
   b. $P_{null} = 0.50$
   c. $P_{real} = 0.51$
   d. all of the above

7. In the case of $N = 3$, $P_{real} = 0.90$, $\alpha = 0.05_{1\ tail}$, using the sign test, the power is _______.

   a. 1.0000
   b. 0.1250
   c. 0.8750
   d. 0.0000

8. Which of the following may be <u>false</u>?

   a. alpha + beta = 1
   b. power + beta = 1
   c. 1 - power = beta

d. 1 - beta = power

9. In a two-tailed test, which of the following would yield the same power as a size of effect represented by $P_{real} = 0.90$?

   a. $P_{real} = 1.00$
   b. $P_{real} = 0.10$
   c. $P_{real} = 0.80$
   d. none of the above

10. If H_0 is false the probability of making a correct decision is ________.

   a. alpha
   b. 1 − alpha
   c. power
   d. 1 − power

11. Assume that you have conducted an experiment and the tail probability for the results you obtained was 0.0900 and $\alpha = 0.05$. You can conclude ________.

   a. the independent variable has no effect
   b. the independent variable has a weak effect
   c. you were unable to detect an effect of the independent variable with your experiment
   d. the null hypothesis is false

12. Truth demands that scientists set _______.

    a. alpha equal to 0.05
    b. alpha equal to 0.01
    c. a or b
    d. not necessarily $\alpha = 0.01$ or 0.05, but use their best judgment

13. In the sign test, if H_0 is false, then P_{real} _______.

    a. equals 0.05
    b. equals 0.50
    c. equals 1 − 0.05
    d. does not equal 0.50

14. In an experiment, if the effect of the independent variable and alpha remain the same and N increases, _______.

    a. power increases
    b. power decreases
    c. power remains the same
    d. cannot be determined

15. In an experiment with $N = 14$ and $P_{real} = 0.10$, what is the power using $\alpha = 0.05_{2\ tail}$?

    a. 0.8417
    b. 0.0132
    c. 0.0000
    d. 0.0500

16. In the same experiment as question 15, what is the power using $\alpha = .01_{2\ tail}$?

    a. 0.0500
    b. 1.0000

c. 0.5847
d. 0.9868

17. An experiment with $N = 18$ is more powerful than an experiment with $N = 17$, all other things being the same.

    a. true
    b. false

18. In order to calculate power you must know _______.

    a. alpha
    b. N
    c. P
    d. the results of the experiment
    e. all of the above
    f. a, b and c

19. In an experiment with $N = 6$ and $P_{real} = 0.15$, what is the power using $\alpha = 0.01_{2\ tail}$?

    a. 1.0000
    b. 0.0500
    c. 0.0000
    d. 0.3771

ANSWERS

Exercises: **1a.** Retain H_0 because 0.1458 > .05 **1b.** 0.0863 **1c.** Type II **2.** 0.1609 **3a.** 0.3552 **3b.** When the effect is in the predicted direction, a one-tailed alpha level results in higher power than a two-tailed alpha level. **4a.** 0.6481 **4b.** 0.8042 **4c.** 0.1958 **4d.** As N increases, power increases **5a.** 0.2968 **5b.** 0.4163 **5c.** 20 subjects **6.** 0.8042 **7a.**

0.0577 **7b.** Increase *N*; increase size of effect. **8.** 0.9423 **9.** False, it is also possible H_0 is true **10.** 0.5400 **11.** 0.2497.

True-False: **1.** T **2.** F **3.** F **4.** T **5.** T **6.** F **7.** F **8.** T **9.** T **10.** T **11.** F **12.** T **13.** F **14.** T **15.** F **16.** T **17.** F.

Self-Quiz: **1.** a **2.** d **3.** a **4.** d **5.** c **6.** b **7.** d **8.** a **9.** b **10.** c **11.** c **12.** d **13.** d **14.** a **15.** a **16.** c **17.** a **18.** f **19.** c.

Chapter 12

Sampling Distributions, Sampling Distribution of the 4Mean, The Normal Deviate (z) Test

CHAPTER OUTLINE

I. Sampling Distribution of a Statistic

A. Definition. a probability distribution of all the possible values of the statistic under the assumption that chance alone is operating.

B. Information given by sampling distributions.

1. All the values that the statistic can take.
2. The probability of getting each value under the assumption that it had resulted due to chance alone.

C. Examples of sampling distributions.

1. Binomial distributions with $P = 0.50$ for the sign test in a replicated measures design.

D. Steps of data analysis.

1. Calculate the appropriate statistic.
2. Evaluate the statistic based on its sampling distribution.

II. Generating Sampling Distributions

A. Derivation of sampling distribution of the statistic.

1. Determine all the possible different samples of size N that can be formed from the population.
2. Calculate the statistic for each of the samples.
3. Calculate the probability of getting each value of the statistic, if chance alone is operating.

B. Null Hypothesis Population.

1. Definition. Actual or theoretical set of population scores that would result if the experiment were done on the entire population and the independent variable had no effect.
2. Use. Used to test the validity of the null hypothesis.

III. The Normal Deviate (z test)

A. Conditions for use.

1. Used when the parameters of the Null Hypothesis Population are known (μ and σ).
2. Uses the mean of the sample as a basic statistic to test the null hypothesis.
3. Must know the sampling distribution of the mean.

IV. Sampling Distribution of the Mean

A. Empirical approach.

1. Draw all possible different samples of a fixed size N from a specific population of raw scores having a mean μ and a standard deviation σ.
2. Calculate the mean of each sample.
3. Calculate the probability of getting each mean value if chance alone were operating.

B. Characteristics of the sampling distribution of the mean.

1. Resulting distribution is a distribution of sample means which itself has a mean and standard deviation.
2. Distribution of sample means is a population set of scores.
3. Mean of the sampling distribution of mean is signified by $\mu_{\overline{X}}$.
4. Standard deviation of the sampling distribution of mean is signified by $\sigma_{\overline{X}}$ and is also known as the standard error of the mean.
5. Mean of sampling distribution equals mean of raw score population:

$$\mu_{\overline{X}} = \mu$$

6. Standard deviation of the sampling distribution of mean is equal to the standard deviation of the raw score population divided by $\sqrt{N}$:

$$\sigma_{\overline{X}} = \frac{\sigma}{\sqrt{N}}$$

where N = the size of the sample.

7. The sampling distribution is normally shaped depending on the shape of the raw score population and

   a. If the raw score distribution is normally shaped, the sampling distribution will be normally shaped.
   b. If the raw score distribution is not normally distributed, the sampling distribution of the mean will be

      1. Normal if N is greater than or equal to 300.
      2. Usually normal for the behavioral sciences if $N \geq$ 30.

V. Computation of z_{obt}

A. Steps.

1. Calculate sample mean, $\overline{X}_{obt}$.
2. Use following formula:

$$z_{obt} = \frac{\overline{X}_{obt} - \mu}{\sigma_{\overline{X}}}$$

where

$$\sigma_{\overline{X}} = \frac{\sigma}{\sqrt{N}}$$

VI. Evaluation of *z* and H_0

A. Definitions.

1. Critical region for rejection of the null hypothesis. The area under the curve that contains all the values of the statistic that allow rejection of H_0.
2. Critical value of a statistic: the value of the statistic that bounds the critical region.

B. Steps.

1. Compute z_{obt}
2. Determine z_{crit} and assess whether z_{obt} falls within the critical region for rejection of H_0. The critical region of rejection is determined by the alpha (α) level.
3. If $|z_{obt}| \geq |z_{crit}|$, reject H_0. If not, retain H_0

C. Conditions for use of the *z* test.

1. Single sample is involved.
2. μ and σ are known.
3. Sampling distribution of mean must be normally distributed, i.e., $N \geq 30$, or the Null Hypothesis Population must be normally distributed.

VII. Power and the z test

A. Main Points.

1. Power is a measure of the sensitivity of the experiment to detect a real effect of the independent variable, if there is one.
2. Defined as the probability of rejecting H_0, if the independent variable has a real effect.

3. Varies directly with N and the magnitude of the independent variable's effect.
4. Varies inversely with α.
5. Power + Beta = 1.

B. Calculation of power with N and μ_{real} given.

1. Determine the critical region for rejection of H_0, using $\overline{X}_{\text{obt}}$ as the statistic.
2. Calculate the probability of getting a sample mean in the critical region for rejection of H_0, assuming sampling is random from the μ_{real} population.

C. Determination of N to achieve a given level of power.

1. Equation.

$$N = \left[\frac{\sigma(z_{\text{crit}} - z_{\text{obt}})}{\mu_{\text{real}} - \mu_{\text{null}}}\right]^2$$

CONCEPT REVIEW

We have been considering the use of the

(1) _______ method to investigate (2) _______. (1) scientific (2) hypotheses

In analyzing our results we want to answer the

question, what is the (3) _______ of getting (3) probability

the obtained results or results even more extreme

if (4) _______ alone is responsible for the
(4) chance

(5) _______ between the experimental
(5) difference

and control scores.

Analyzing the results of an experiment involves two steps. First, calculating the appropriate

(6) _______, and second, evaluating the
(6) statistic

statistic based on the appropriate (7) _______
(7) sampling

(8) _______.
(8) distribution

The sampling distribution of a statistic gives two important pieces of information. First, it gives

all the (9) _______ that the (10) _______ can
(9) values’
(10) statistic

take. Second, it gives the (11) _______ of
(11) probability

getting each value under the (12) _______ that
(12) assumption

it had resulted due to (13) _______ alone.
(13) chance

For example, with the (14) _______
(14) replicated

measures design we used the sign test to analyze

the data. The (15) _______ calculated was
(15) statistic

the (16) _______ of pluses in the sample of N
(16) number

(17) _______ scores. The results were analyzed	(17) difference
using the (18) _______ (19) _______ with	(18) binomial (19) distribution
$P = 0.50$ as the appropriate (20) _______	(20) sampling
(21) _______. In this case there is a	(21) distribution
(22) _______ sampling distribution with each	(22) different
different value of (23) _______.	(23) N
To draw our conclusions for any experiment	
we compare the (24) _______ of	(24) probability
the (25) _______ statistic or one more	(25) obtained
extreme to the (26) _______ (27) _______.	(26) alpha (27) level
If the probability of the obtained statistic is	
(28) _______ than or (29) _______	(28) less (29) equal
to alpha we (30) _______ the (31) _______	(30) reject (31) null
(32) _______. If not, we (33) _______ H_0.	(32) hypothesis (33) retain
This strategy is the same for all experiments.	
What varies is the (34) _______ used and its	(34) statistic

accompanying (35) _______ (36) _______. (35) sampling. (36) distribution

Using the empirical approach, sampling distributions of a statistic are derived by first determining all the (37) _______ different samples of size (38) _______ that can be formed from the population; second, calculating the (39) _______ for each of these (40) _______; and third, calculating the (41) _______ of getting each value if (42) _______ alone is operating. This is done using a theoretical set of scores which would result if the independent variable had no effect. This population set of scores is called the (43) _______ (44) _______ (45) _______.

(37) possible
(38) *N*
(39) statistic
(40) samples
(41) probability
(42) chance
(43) Null
(44) Hypothesis
(45) Population

More formally, the Null Hypothesis Population is an actual or theoretical set of population scores which would result if the experiment were done on the entire (46) _______ and the

(46) population

(47) _______ variable had (48) _______ effect.

(47) independent
(48) no

It is called the Null Hypothesis Population

because it is used to test the (49) _______

(49) validity

of the (50) _______(51) _______. A sampling

(50) null
(51) hypothesis

distribution gives all the values a statistic can

take, along with the probability of getting each

value if (52) _______ is (53) _______ from the

(52) sampling
(53) random

Null Hypothesis Population.

The Normal Deviate

The (54) _______ test or (55) _______

(54) *z*
(55) normal

(56) _______ test is used when we

(56) deviate

know the (57) _______ of the Null Hypothesis Population. The statistic used is the

(57) parameters

(58) _______ of the sample. Therefore,

(58) mean

the appropriate probability distribution to

use is the sampling distribution of the

(59) _______. This sampling distribution is | (59) mean

constructed by taking a specific population

of (60) _______ scores having a mean of | (60) raw

(61) _______ and standard deviation of | (61) μ

(62) _______. Then one draws all | (62) σ

possible different (63) _______ of size | (63) samples

(64) _______, calculates the (65) _______ | (64) N (65) mean

of each sample, and calculates the

(66) _______ of getting each mean value | (66) probability

if chance alone were operating. The result

is the sampling distribution of the mean

for samples of size (67) _______ for the specific | (67) N

population parameters (68) _______ and | (68) μ

(69) _______. | (69) σ

The sampling distribution of the mean for a

given size N has features of its own, namely a

(70) _______ and (71) _______ | (70) mean (71) standard

(72) _______. The mean of the sampling distribution of the mean is symbolized by (73) _______. The standard deviation of the sampling distribution of the mean is symbolized by (74) _______. This standard deviation is also known as the (75) _______ (76) _______ of the mean. The sampling Distribution of the mean has a mean equal to the (77) _______ of the (78) _______score population. In equation form: $\mu_{\overline{X}} = \mu$

It also has a standard deviation equal to the (79) _______(80) _______ of the (81) _______score population divided by (82) _______. In equation form:

$$(83) \_\_\_\_\_ = \frac{\sigma}{\sqrt{N}}$$

The sampling distribution of the mean is

(72) deviation
(73) $\mu_{\overline{X}}$
(74) $\sigma_{\overline{X}}$
(75) standard
(76) error
(77) mean
(78) raw
(79) standard
(80) deviation
(81) raw
(82) $\sqrt{N}$
(83) $\sigma_{\overline{X}}$

(84) _______ shaped depending on the raw score population and on the (85) _______ (86) _______. If the shape of the raw score population is normally distributed, then the sampling distribution will be (87) _______ distributed (88) _______ of sample size. If the raw scores are not normally distributed, then the sampling distribution of the mean becomes more (89) _______ as *N* (90) _______.

(84) normally
(85) sample
(86) size
(87) normally
(88) regardless
(89) normal
(90) increases

If N is greater than (91) _______ then the sampling distribution is almost (92) _______ normally shaped. In the behavioral sciences if *N* is greater than or equal to (93) _______, it is usually assumed that the sampling distribution of the mean is normally shaped.

(91) 300
(92) always
(93) 30

The *z* equation for sample means is very similar to the *z* equation for single scores. The equation for the *z* test for sample means is:

$$z_{\text{obt}} = \frac{(94)\,\_\_\_\_\_ - (95)\,\_\_\_\_\_}{(96)\,\_\_\_\_\_}$$

(94) $\overline{X}_{obt}$
(95) $\mu_{\overline{X}}$ or μ
(96) $\sigma_{\overline{X}}$

One can then compare the (97) _______ of Getting (98) _______ with the (99) _______ (100) _______ to draw conclusions about the experiment. Another way to analyze the results is to determine the (101) _______ (102) _______ for rejection of the null hypothesis. This is the (103) _______ under the curve that contains all the (104) _______ of the (105) _______ that will allow rejection of the null hypothesis.

(97) probability
(98) z_{obt}
(99) alpha
(100) level
(101) critical
(102) region
(103) area
(104) values
(105) statistic

The (106) _______ (107) _______ of a statistic is the value of the statistic that (108) _______ the critical region. The critical region is determined by the

(106) critical
(107) value
(108) bounds

(109) _______(110) _______. The combined	(109) alpha
area under the critical region	(110) level
must (111) _______ the alpha level. If |	(111) equal
(112) _______| ≥	(112) z_{obt}
|(113) _______|, reject he null hypothesis.	(113) z_{crit}
If not, (114) _______ the null hypothesis.	(114) retain
In order to use the z test there must be a	
(115) _______ sample and the parameters	(115) single
of the Null Hypothesis Population,	
(116) _______ and (117) _______ must	(116) μ (117) σ
be known. Also, the sampling distribution of	
the mean must be (118) _______	(118) normally
(119) _______. This means *N* must be greater	(119) distributed
than or equal to (120) _______	(120) 30
or the Null Hypothesis Population must be	
normally distributed.	
The power of a test or of an experiment	
is a measure of the (121) _______ of the	(121) sensitivity

experiment to the real effect of the independent	
variable. Thus, power is defined as the	
probability of (122) _______ H_0 when	(122) rejecting
the independent variable has a	
real effect in the experiment. It is also the	
probability of rejecting H_0 when	
(123) _______ is false. If the real effect	(123) H_0
of the independent variable increases,	
power (124) _______. If the number of	(124) increases
subjects in an experiment is increased, with	
other factors held constant, power also	
(125) _______. There is an (126) _______	(125) increases (126) inverse
relationship between power and beta. The	
(127) _______ the power, the less the chance	(127) higher
of a Type II error. Since making alpha more	
stringent, causes beta to (128) _______,	(128) increase
making alpha more stringent, will cause	
power to (129) _______. In order to minimize	(129) decrease

Type I and II errors, it is desirable to make

alpha (130) _______ and power | (130) low

(131) _______. For a given low (stringent) | (131) high

alpha level, power can be increased by

increasing (132) _______ and increasing | (132) *N*

the (133) _______ of the (134) _______ | (133) effect
(134) independent

(135) _______. | (135) variable

EXERCISES

For the following problems assume the Null Hypothesis Populations are normally distributed.

1. What is the critical value for *z* for each of the following alpha levels?

   a. 0.05 2 tail
   b. 0.01 2 tail
   c. 0.05 1 tail
   d. 0.01 1 tail
   e. 0.02 2 tail

2. What is the value of $\mu_{\overline{X}}$ for the following values of μ?

   a. 15.2
   b. 14.0
   c. 10.0

3. List all the possible values of $\overline{X}$ for samples of size $N = 2$ taken from the population of scores 0, 1, 2 (with replacement).

4. For the means of size $N = 2$ in problem 3, what is the value for:

   a. $\mu_{\overline{X}}$
   b. $\sigma_{\overline{X}}$

5. Assume the following values of σ have been obtained for different Null Hypothesis Populations. Calculate the value of the standard error of the mean for samples of size N taken from the respective populations.

   a. $\sigma = 26.2$, $N = 9$
   b. $\sigma = 5.6$, $N = 2$
   c. $\sigma = 13.0$, $N = 6$
   d. $\sigma = 1000$, $N = 100$
   e. $\sigma = 10.0$, $N = 3$

6. In a population with $\mu = 100$ and $\sigma = 15$, what is the probability if we randomly drew a sample of $N = 9$ we would get a value of $\overline{X} \geq 108$?

7. In problem 6, if you were testing the hypothesis that the sample mean (108) was drawn from the Null Hypothesis

Population with $\mu = 100$ and $\sigma = 15$, what would you conclude using $\alpha = 0.05_{1\ \text{tail}}$?

8. You have just read that chickens have a relatively low level of body fat. The population mean percentage of fat in a chicken is reported to be 32% with the standard deviation of 8.7%. How likely is it that you could randomly select a sample of 12 chickens from the population that would have a mean of 26% body fat or less?

9. A computer memory manufacturer specifies that its memory chip stores data incorrectly an average of 6.3 out of 10 million cycles with a standard deviation of 0.48. A batch of 30 chips your company ordered stores data incorrectly an average of 6.9 times per 10 million cycles.

   a. Does it seem reasonable that your 30 chips are a random sample from a population with the specifications given by the computer memory manufacturer? Use $\alpha = 0.05_{2\ \text{tail}}$.
   b. What is the critical value of z?

10. The average weight of a species of laboratory rats at birth is 27.6 grams with a standard deviation of 3.4 grams. You wish to test the hypothesis that a different maternal feeding cycle but the same quantity and quality will affect birth-weight. A sample of 10 female rats given the new feeding schedule gave birth to 56 baby rat pups with an average weight of 25.9 grams.

    a. What do you conclude using $\alpha = 0.01_{2\ \text{tail}}$?
    b. If the real effect of the new feeding schedule is to cause an average weight at birth of 24.5 grams, what is the power of the experiment to detect this effect?
    c. If the sample size was increased to 25, what is the power of the experiment to detect the real effect given in 10b?

d. What size of sample would it be necessary to use to have power =.9000 to detect an effect which produces an average birth weight of 26.0?

11. From a population of 4 different scores (e.g. 3, 5, 7, 9) how many samples of size $N = 2$ are possible sampling one at a time with replacement? (Solve by enumeration.)

12. The average life of a light bulb is 862 hours with a standard deviation of 51 hours. A new manufacturing process results in a sample of 24 bulbs with a mean life of 899 hours until burn out.

    a. Would you recommend a change in manufacturing technique using $\alpha = 0.01_{1\ \text{tail}}$?
    b. What type error might you be making?
    c. If the new manufacturing process has an effect so as to produce an average life of 880 hours, what is the power of the experiment to detect this effect?
    d. What should the sample size be to have a power =0.8000 to detect the real effect postulated in 12c?
    e. What effect does making alpha less stringent have on power? Illustrate by recomputing power for question 12c, only this time use $\alpha = 0.05_{1\ \text{tail}}$.

13. Using $\alpha = 0.01_{2\ \text{tail}}$, test the hypothesis that the following sample could have been drawn from a population with $\mu =$ 27.2 and $\sigma = 2.1$.

X: 25.2, 29.9, 24.8, 26.0, 22.1

TRUE FALSE QUESTIONS

T F 1. The binomlal distribution is an example of a sampling distribution.

T F 2. A sampling distribution is based on the assumption that the independent variable has an effect on the dependent variable.

T F 3. Only one sampling distribution can apply to any given problem or experiment.

T F 4. The sampling distribution of the mean changes as sample size changes.

T F 5. A sampling distribution gives one the values a statistic can take, along with the probability of getting each value if sampling is random from the H_1 population.

T F 6. To use the normal deviate one must know the population parameters of the Null Hypothesis Population.

T F 7. If the sample mean is different from the population mean, H_0 must be false.

T F 8. $\sigma_{\overline{X}}$ is sometimes called the standard error of the mean because each sample can be considered an estimate of the mean of the raw score population and variability between sample means occur due to errors in estimation.

T F 9. For sampling distribution of the mean, $\mu_{\overline{X}} = \mu$

T F 10. As N increases, $\sigma_{\overline{X}}$ decreases.

T F 11. In general, if $N \geq 30$, the sampling distribution of the mean will be normally shaped.

T F 12. For the population 2, 4, 6, 8, 10, $\mu_{\overline{X}} = \frac{10}{\sqrt{5}}$.

T F 13. The formula for z_{obt} for the sampling distribution of the mean is $z_{\text{obt}} = \frac{X - \mu}{\sigma}$.

T F 14. If z_{obt} falls within the critical region for rejection of H_0, one can be certain H_0 is in reality false.

T F 15. If $\alpha = 0.05_{2\ \text{tail}}$, the critical value of z is ±1.96.

T F 16. The z test is used to test whether or not it is reasonable to assume a sample of size N was likely to have been drawn from a population with known parameters μ and σ.

T F 17. For the sampling distribution of the mean, $\sigma_{\overline{X}} = \sigma$.

T F 18. $\alpha + \beta = 1.00$.

T F 19. If there is a real effect, the higher the power, the more likely H_0 will be rejected.

T F 20. Increasing N usually results in an increase of power.

T F 21. If H_0 is false, but power is low, there is a high probability we will err in our conclusion.

T F 22. By using a stringent alpha level, and designing the experiment for high power, we maximize the probability of correctly concluding regardless of whether H_0 is true or false.

T F 23. Power is independent of the magnitude of the independent variable's effect.

T F 24. Power is an esoteric topic that has very little practical utility.

SELF-QUIZ

1. A raw score distribution that is negatively skewed will produce a sampling distribution of the mean for $N = 4$ which is _______.
   a. also negatively skewed
   b. normally distributed
   c. positively skewed
   d. cannot be determined

2. A raw score distribution which has a moderate negative skew will result in a sampling distribution of the mean for $N = 42$ that is _______.

   a. also negatively distributed
   b. normally distributed
   c. positively skewed
   d. cannot be determined

3. If one draws all possible samples for various values of N from the same population of raw scores, as N increases _______.

a. the standard error of the mean increases
b. the standard error of the mean stays the same
c. the standard error of the mean decreases
d. the standard error of the mean cannot be calculated

4. If one draws all possible samples for various values of N from the same population of raw scores, as N increases _______.

a. the mean of the sampling distribution of the mean increases
b. the mean of the sampling distribution of the mean stays the same
c. the mean of the sampling distribution of the mean decreases
d. none of the above

5. In cases where $N > 1$, the relationship between the raw score population standard deviation and the standard error is _______.

a. the standard error is greater than the standard deviation
b. the standard error is less than the standard deviation
c. the standard error equals the standard deviation
d. the standard error is the standard deviation

6. The area under the critical region for rejection must _______.

a. equal alpha
b. be less than alpha
c. be greater than alpha
d. equal one-half of alpha

7. The critical value of the statistic _______.

a. is independent of alpha
b. is equal to alpha

c. is always the same in the z distribution
d. depends on alpha

8. The critical value for the z distribution using $\alpha = 0.05$ 2 tail is ________.

a. ±1.96
b. ±2.58
c. ±1.64
d. ±2.33

9. The critical value for the z distribution using $\alpha = 0.01$ 2 tail is ________.

a. ±1.96
b. ±2.58
c. ±1.64
d. ±2.33

10. Assuming that the population mean is 47.2 and the population deviation is 6.4, what is the z_{obt} value for a sample mean of 52.1 if $N = 8$?

a. 1.96
b. 2.17
c. 1.73
d. 0.77

11. If z_{obt} was 2.60 and z_{crit} was 2.58, one would ________.

a. accept H_0
b. reject the alternative hypothesis
c. retain H_0
d. reject H_0

12. To use the z test the data _______.

    a. must be normally distributed
    b. must be sampled from a normal distribution
    c. must be sampled from a normal distribution if sample size is 10
    d. can be any shape if $N = 10$

13. What would one conclude about the null hypothesis that a sample of $N = 46$ with a mean of $\overline{X} = 104$ could reasonably have been drawn from a population with the parameters of $\mu = 100$ and $\sigma = 8$? Use $\alpha = 0.05_{2\text{ tail}}$.

    a. accept H_0
    b. reject H_1
    c. reject H_0
    d. retain H_0

14. If the population standard deviation is 19.0 and the sample size is 19, then the standard error equals _______.

    a. 19.00
    b. 1.00
    c. 4.36
    d. cannot be determined; need to know population mean

15. If one draws two samples of size N from a distribution of raw scores (with replacement), then the means of the two samples will _______.

    a. be equal
    b. perhaps be equal but not necessarily
    c. be equal to the standard error
    d. be equal to 1.0

16. What we are testing with the *z* test in this chapter is _______.

    a. the sample mean
    b. the population mean
    c. the sample error
    d. the standard error

17. Could the following sample reasonably have been drawn from a normal population with a mean of 20 and standard deviation of 1.5 using $\alpha = 0.01_{2\text{ tail}}$?

    ***X*: 21, 21, 21, 20, 22, 20, 22**?

    a. yes
    b. no
    c. cannot be tested with *z* test
    d. insufficient information

18. A health psychologist knows the smoking population at the hospital where she works smokes an average of 18 cigarettes per day with a standard deviation of 7. She plans to conduct a program to reduce smoking. If she has 25 persons in her program, what is the power to detect a real effect of the program such that cigarette consumption is reduced, on the average, by 5 cigarettes? Assume population normality and use $\alpha = 0.05_{1\text{ tail}}$.

    a. Insufficient information
    b. 0.9732
    c. 0.9463
    d. 0.8925

ANSWERS

Exercises: **1a.** ±1.96 **1b.** ±2.58 **1c.** 1.64(5) **1d.** 2.33 **1e.** ±2.33 **2a.** 15.2 **2b**. 14.0 **2c**. 10.0 **3.** 0, 0.5, 1, 1.5, 2 **4a.** 1.0 **4b.** .5773 **5a.** 8.73 **5b.** 3.96 5**c.** 5.31 **5d.** 100.00 **5e.** 5.77 **6.** 0.0548 **7.** Retain H_0 **8.** p = 0.0084 **9a.** z_{obt} = 6.85, therefore, reject H_0. Conclude that the sample of 30 chips with a mean of 6.9 failures could not reasonably have been drawn from the Null Hypothesis Population. **9b.** ±1.96 **10a.** z_{obt} = −1.58 and z_{crit} = ±2.58. Retain H_0. **10b.** Power = 0.6179 **10c.** Power = 0.9761 **10d.** N = 68 **11.** 16 **12a.** Reject H_0 (z_{obt} = 3.55, z_{crit} = 2.33). Assuming there were no economic disadvantages in changing manufacturing processes, it appears as if this process should be adopted. **12b.** Type I **12c.** Power = 0.2743 **12d.** N = 81 **12e.** It increases power. Power = 0.5319 **13.** Retain H_0 (z_{obt} = −1.70, z_{crit} = ±2.58).

True-False: **1.** T **2.** F **3.** F **4.** T **5.** F **6.** T **7.** F **8.** T **9.** T **10.** T **11.** T **12.** F **13.** F **14.** F **15.** T **16.** T **17.** F **18.** F **19.** T **20.** T **21.** T **22.** T **23.** F **24.** F.

Self-Quiz: **1.** a **2.** b **3.** c **4.** b **5.** b **6.** a **7.** d **8.** a **9.** b **10.** b **11.** d **12.** c **13.** c **14.** c **15.** b **16. a** **17.** a **18.** b

Chapter 13

Student's t Test for Single Samples

CHAPTER OUTLINE

I. Introduction

A. Use of the *t* test. Use the *t* test when

1. the mean of the Null Hypothesis Population can be specified
2. the standard deviation is unknown (which also means the *z* test cannot be used)

B. Applications covered in this chapter.

1. Analysis of data in experiments with a single sample
2. Determining the significance of Pearson *r*.

II. Comparison of *z* and *t* Tests

A. Formula.

$$z_{obt} = \frac{\overline{X}_{obt} - \mu}{\sigma_{\overline{X}}} \quad \text{where } \sigma_{\overline{X}} = \frac{\sigma}{\sqrt{N}}$$

$$t_{obt} = \frac{\overline{X}_{obt} - \mu}{s_{\overline{X}}} \quad \text{where } s_{\overline{X}} = \text{estimated standard error of the mean} = \frac{s}{\sqrt{N}}$$

B. <u>*z* and *t* equation differences</u>. Difference between equations is that in the *t* equation, s and $s_{\overline{X}}$ are used instead of σ and $\sigma_{\overline{X}}$ respectively.

III. Sampling Distribution of *t*

A. <u>Definition</u>. the sampling distribution of *t* is a probability distribution of the *t* values which would occur if all possible different samples of a fixed size *N* were drawn from the Null Hypothesis Population. It gives (1) all the possible *t* values for samples of size *N* and (2) the probability of getting each value if sampling is random from the Null Hypothesis Population.

B. <u>Characteristics</u>.

1. Family of curves; many curves with a different curve for each sample size *N*.
2. Shape. Shaped similarly to *z* distribution if

   a. sample size ≥ 30 or
   b. H_0 population is normally distributed.

C. Degrees <u>of freedom (df)</u>. The number of scores that are free to vary.

$$\mathbf{df = N - 1}$$

D. <u>Comparison of *z* and *t* distributions</u>.

1. *t* and *z* are both symmetrical about 0
2. As df increases *t* becomes more similar to *z*
3. As df approaches ∞, *t* becomes identical to *z*
4. At any value of df $< \infty$, the *t* distribution has more extreme values than *z* (i.e., tails of the *t* distribution are more elevated than in the *z* distribution)
5. For a given alpha level, $t_{crit} > z_{crit}$

IV. Calculations and Use of *t*

A. <u>Calculation of *t* from raw scores</u>.

$$t_{obt} = \frac{\overline{X}_{obt} - \mu}{\sqrt{\dfrac{SS}{N(N-1)}}}$$

B. <u>Requirements</u>. Appropriate use of *t* requires that sampling distribution of $\overline{X}$ is normal. This can result if $N \geq 30$ or population of raw scores is normal.

V. Size of Effect Using Cohen's *d*

A. <u>Rationale</u>. The statistic we are using to measure size of effect is symbolized by "*d*." It is a standardized statistic that relies on the relationship between the size of effect and $|\overline{X}_{obt} - \mu|$. As the size of effect gets greater, so does $|\overline{X}_{obt} - \mu|$, regardless of the direction of the effect. The statistic *d* uses the absolute value of $\overline{X}_{obt} - \mu$ since we are interested in the *size* of real effect, and are not concerned about direction. This allows *d* to have a positive value that increases with the size of the difference between $\overline{X}_{\text{obt}}$ and regardless of the direction of the real effect. $\left|\overline{X}_{\text{obt}} - \mu\right|$ is divided by to create a standardized value, much as was done with *z* scores.

B. <u>Formula for Cohen's *d*.</u>

$$d = \frac{\left|\overline{X}_{\text{obt}} - \mu\right|}{\sigma}$$

Conceptual equation for size of effect, single sample t test

Since is unknown, we estimate it using *s*, the sample standard deviation. Substituting *s* for , we arrive at the computational equation for size of effect. Since *s* is an estimate, $\hat{d}$ is used instead of *d*.

$$\hat{d} = \frac{\left|\overline{X}_{\text{obt}} - \mu\right|}{s}$$

Computational equation for size of effect, single sample t test

C. <u>Interpreting the Value of $\hat{d}$</u>. To interpret the value of $\hat{d}$, we are using the criteria that Cohen has provided. These criteria are given in the following table.

Value of $\hat{d}$	Interpretation of $\hat{d}$
0.00 – 0.20	Small effect
0.21 – 0.79	Medium effect
≥0.80	Large effect

VI. Confidence Intervals for the Population Mean

A. Definition. A confidence interval is a range of values which probably contains the population mean. Confidence limits are the values that bound the confidence interval. Example: The 95% confidence interval is an interval such that the probability is 0.95 that the interval contains the population value.

B. Formula for confidence interval.

$$\mu_{\text{lower}} = \overline{X}_{\text{obt}} - s_{\overline{X}} t_{\text{crit}}$$

$$\mu_{\text{upper}} = \overline{X}_{\text{obt}} + s_{\overline{X}} t_{\text{crit}}$$

VII. Testing Significance of Pearson *r*.

A. Rho (ρ). this is the Greek letter to symbolize the population correlation coefficient.

B. Nondirectional H_0. Asserts $\rho \neq 0$.

C. Directional H_0 Asserts that ρ is positive or negative depending on the predicted direction of the relationship.

D. Sampling distribution of *r*. Generated by taking all samples of size *N* from a population in which $\rho = 0$ and calculating *r* for each sample. By systematically varying the population scores and *N*, the sampling distribution of *r* is generated.

E. Using *t* test to evaluate significance of *r*:

$$t_{\text{obt}} = \frac{r_{\text{obt}} - \rho}{s_r} \quad \text{where} \quad s_r = \frac{1 - r_{\text{obt}}^{2}}{N - 2}$$

with df = $N - 2$ and N equals the number of pairs of X, Y scores.

F. <u>Using r_{crit} to evaluate the significance of r:</u>

if $|r_{obt}| \geq |r_{crit}|$, reject H_0

The values of r_{crit} can be calculated directly. Values of r_{crit} are shown in Table E.

CONCEPT REVIEW

The (1) _______ test is appropriate in	(1) z
situations where both the (2) _______ and	(2) mean
(3) _______ of the Null Hypothesis Population	(3) standard
are known. However, it is more common that	
we can specify deviation the mean of the Null	
Hypothesis Population and the standard	
deviation is unknown. In this case we can use	
the (4) _______ test. The formula for the t test is:	(4) t

$$t_{obt} = \frac{(5)\_\_\_\_ - (6)\_\_\_\_}{(7)\_\_\_\_}$$

(5) $\overline{X}$
(6) μ
(7) $s_{\overline{X}}$

where

$$s_{\overline{X}} = \frac{(8)\_\_\_\_}{\sqrt{(9)\_\_\_\_}}$$

(8) *s*
(9) *N*

This equation differs from the equation for z_{obt}

only in that (10) _______ is used instead of (10) *s*

(11) _______ in the denominator. (11) σ

When σ is unknown, we (12) _______ it using (12) estimate

the estimate given by (13) _______. When σ is (13) *s*

estimated, the resulting statistic is called

(14) _______. The (15) _______ (14) *t* (15) sampling

(16) _______ of *t* is a probability distribution of (16) distribution

the *t* values that would occur if all possible

different samples of a fixed size were drawn

from the (17) _______ (18) _______ (17) Null (18) Hypothesis

(19) _______. It gives all the possible values (19) Population

of (20) _______ for samples of size *N* and (20) *t*

the (21) _______ of getting each value if (21) probability

sampling is (22) _______ from the Null Hypothesis Population. If the Null Hypothesis Population is (23) _______ (24) _______ or N (25) _______ 30, then the t distribution looks very much like the (26) _______ distribution, except that there is a (27) _______ of curves that vary with (28) _______ (29) _______.

(22) random
(23) normally
(24) shaped
(25) ≥
(26) z
(27) family
(28) sample
(29) size

The t distribution varies uniquely with (30) _______ of (31) _______. The degrees of freedom, symbolized (32) _______, for any statistic is the (33) _______ of scores that are free to (34) _______ in calculating that (35) _______. For the t test for single samples there are (36) _______ df.

(30) degrees
(31) freedom
(32) df
(33) number
(34) vary
(35) statistic
(36) $N - 1$

The t distribution is (37) _______ about zero. When df approaches (38) _______ the t distribution becomes (39) _______ to

(37) symmetrical
(38) infinity
(39) identical

the *z* distribution. At any number of df less

than infinity the distribution has (40) _______ (40) more

variability than the *z* distribution. That is, the

(41) _______ of the *t* distribution (41) tail

is higher than the *z* distribution. For a given

alpha level, the critical value of *t* is (42) _______ (42) higher

than for *z*. This is because we use (43) _______ (43) *s*

o estimate (44) _______ and we cannot be (44) σ

certain that our estimate is perfectly accurate.

To calculate the value of t_{obt} from raw data

we can use the following formula.

$$t_{obt} = \frac{\overline{X}_{obt} - \mu}{\sqrt{\frac{(45)\_\_\_\_\_}{(46)\_\_\_\_\_ \times [(47)\_\_\_\_\_ - (48)\_\_\_\_\_]}}}$$

(45) *SS*
(46) *N*
(47) *N*
(48) 1

Where (49) _______ $= \Sigma X^2 - (\Sigma X)^2/N$ (49) *SS*

To use the *t* test, the sampling distribution

of (50) _______ must be (51) _______. (50) $\overline{X}$
(51) normal

Size of Effect Using Cohen's *d*

Cohen's *d* uses the absolute value of

(52) _______ - (53) _______ to measure the | (52) $\overline{X}_{\text{obt}}$
(53)

size of effect, because the size of effect

varies (54) _______ with this absolute value. | (54) directly

Rather than using raw score units, *d* is a

(55) _______ value achieved by dividing | (55) standardized

$|\overline{X}_{\text{obt}} - \mu|$ by (56) _______. The conceptual | (56)

formula for *d* is given by

$$d = \frac{|(57)\_\_\_\_\_ - (58)\_\_\_\_\_|}{(59)\_\_\_\_\_}$$

(57) $\overline{X}_{\text{obt}}$
(58)
(59)

Since is unknown, we estimate it with

(60) _______, resulting in the following | (60) *s*

computational equation for size of effect.

$$\hat{d} = \frac{|(61)\_\_\_\_\_ - (62)\_\_\_\_\_|}{(63)\_\_\_\_\_}$$

(61) $\overline{X}_{\text{obt}}$
(62)
(63) *s*

Confidence Intervals

Unless we have sampled the (64) _______ (64) entire

population we cannot know the population

mean exactly. We use the (65) _______ (65) sample

mean as an (66) _______ of the true population (66) estimate

mean. When only one value is used as an

estimate this is called a (67) _______ (67) point

(68) _______. In estimating the population (68) estimate

mean, the usual way is to give a (69) _______ (69) range

of values for which one is reasonably

(70) _______ that the range (70) confident
(71) _______ the population mean. This is (71) includes

Called (72) ______ (73) _______. (72) interval
(73) estimation

A (74) _______ interval is a range of (74) confidence

(75) _____ which probably contains the (75) values

population mean. Confidence (76) _______ (76) limits

are the values that (77) _______ the (77) bound

confidence interval. The 95% confidence interval

is an interval such that (78) _______ is (78) probability

(79) _______ that the interval contains the population value. There is a probability of	(79) 0.95
(80) _______ that such an interval will not contain the population value. The general formula for the confidence interval is	(80) 0.05
μ_{lower} = (81) _______ – (82) _______ x	(81) $\overline{X}_{\text{obt}}$ (82) $s_{\overline{X}}$
(83) _______	(83) t_{crit}
μ_{upper} = (84) _______ + (85) _______	(84) $\overline{X}_{\text{obt}}$ (85) $s_{\overline{X}}$
x (86) _______	(86) t_{crit}
where t_{crit} = the critical (87) _______ - tailed value of *t* corres-ponding to the desired confidence interval.	(87) one
The (88) _______ confidence interval will always be larger (wider) than the	(88) 99%
(89) _______ confidence interval. The greater the interval, the more (90) _______ we have that it contains the population mean.	(89) 95% (90) confidence

Significance of *r*

To determine whether a correlation exists in the population, we must test the (91) _______ of the obtained (92) _______.

(91) significance
(92) r

The population correlation coefficient is symbolized by the Greek letter (93) _______.

(93) ρ (rho)

The nondirectional alternative hypothesis asserts that ρ = (94) _______. The null hypothesis is tested by assuming the sample set of X and Y scores, which have a correlation equal to (95) _______, is drawn from a population where (96) _______ = 0. Using the t test, the significance of r can be evaluated using the formula:

(94) 0
(95) r_{obt}
(96) ρ

$$t_{obt} = \frac{(97)\ \_\_\_\_\_ - \rho}{(98)\ \_\_\_\_\_}$$

(97) r_{obt}
(98) s_r

Where

$$s_r = \frac{1-(99)\ \_\_\_\_\_}{(100)\ \_\_\_\_\_ - 2}$$

(99) r_{obt}
(100) N

with df = (101) _______ – (102) _______.
(101) N
(102) 2

N is the number of (103) _______ of scores
(103) pairs

and s_r is the (104) _______ of the
(104) estimate

standard deviation of the sampling distribution

of (105) _______ Again, this is a t test because
(105) r

we are (106) _______ the standard
(106) estimating

deviation of the (107) _______ (108) _______.
(107) sampling
(108) distribution

A table of r_{crit} can be made by substituting

(109) _______ into the above equation and
(109) t_{crit}

solving for (110) _______ for any level of
(110) r_{crit}

(111) _______ and (112) _______. Then we
(111) α
(112) df

can apply the decision rule:

If |(113) _______ | ≥ |(114) _______
(113) r_{obt}
(114) r_{crit}

|, reject H_0

EXERCISES

For the following problems, assume the Null Hypothesis Populations are normally distributed.

1. What are the values for the degrees of freedom for samples of the following sizes (assuming one is calculating a t test for a single sample)?

   a. 25
   b. 24
   c. 2
   d. 11
   e. 9

2. A sample has a value of $\overline{X}_{obt}$ = 46, s = 8, and N = 12.

   a. What is the value of tobt to test the hypothesis that this sample could reasonably have been drawn from the Null Hypothesis Population with μ = 50?
   b. What is the critical value of t using □ = 0.052 tail?
   c. What do you conclude?

3. What are the critical values of t for each of the following values of N and alpha using a nondirectional hypothesis?

N	□
a. 12	0.05
b. 20	0.01
c. 2	0.05
d. 5	0.02
e. 19	0.01

Now using a directional hypothesis?

N	□
f. 13	0.025
g. 17	0.005
h. 8	0.05
i. 15	0.01
j. 10	0.05

4. One wishes to investigate the hypothesis that exercise reduces systolic blood pressure. The population mean for the systolic blood pressure of people who do not exercise is 120 mmHg. A sample of 13 members of a local running club has a mean systolic blood pressure of 113 mmHg with a standard deviation of 9.2 mmHg.

   a. State H1 (directional)
   b. State H0
   c. What do you conclude using □ = .011 tail
   d. What is the size of the effect?

5. Using □ = 0.052 tail, is it reasonable to assume that the following set of scores could have been randomly drawn from a population with a mean of 74.5?

   ***X*: 73.0, 72.1, 72.0, 69.1, 70.8**

6. The average growth per day of a certain type lawn seed using standard fertilizer is 2.6 inches per week. You wish to test the effects of a fertilizer on growth rate. You add fertilizer to several different patches of grass grown from that seed and observe the following results:

Growth per Week (inches)
1.9
2.0
3.5
4.2
4.0
3.2
2.8

a. What is the value of tobt?
b. What do you conclude using □ = 0.052 tail?

7. You believe that sleep affects memory. Having collected data on a large number of subjects who are getting ordinary sleep, you know that the mean digit span of this population is 5.6 digits. To test your hypothesis you measure the digit span of 15 volunteers randomly selected from the population, after depriving them of one night's sleep. The results show $\overline{X}_{obt}$ = 4.1 and s = 1.4.

   a. What do you conclude, using □ = 0.052 tail?
   b. What is the size of effect?

8. Consider the following set of scores:

   102, 102, 106, 105, 104, 104, 107, 108

   a. What is the 95% confidence interval for the population mean?
   b. What is the 99% confidence interval for the population mean?

9. If $\overline{X}$ = 50 and s = 6.7:

a. If N = 18, what is the 95% confidence interval for the population mean?
b. If N = 18, what is the 99% confidence interval for the population mean?
c. If N = 10, what is the 95% confidence interval for the population mean?
d. If N = 10, what is the 99% confidence interval for the population mean?

10. Assume that μ = 500 and □ = 100. Your study shows a sample of size 22 with a mean of 530 and standard deviation of 113.

a. What is the most powerful test to use to test the hypothesis that the mean of the sample was drawn from the above Null Hypothesis Population?
b. What is the value of the test statistic?
c. What do you conclude using □ = 0.052 tail?

11. Using the information in problem 10:

a. What is the value of tobt?
b. Does z or t allow one to reject H0 more easily?
c. What type error might one be making in this problem?

12. Assume that you have just calculated the correlation coefficient from an experiment with N = 16 pairs of observations. The value of robt is 0.493.

a. What is the value of tobt for testing the hypothesis that r is significantly different from zero?
b. What do you conclude using □ = 0.052 tail?
c. What is the value of rcrit?

13. Consider the following set of scores:

X	10	15	20	2	5	30	35
Y	11	12	16	1	3	19	20

a. What is the value of the correlation coefficient?
b. What is the value of rcrit using □ = 0.052 tail?
c. Is robt significantly different from zero using □ = 0.052 tail?

TRUE-FALSE QUESTIONS

T F 1. The *t* test for single samples is used when the population parameters μ and σ are unknown.

T F 2. In the *t* test *s* is used to estimate σ.

T F 3. In general t_{crit} is greater than z_{crit} at the same α level.

T F 4. The *t* test is more powerful than the *z* test.

T F 5. The *t* distribution is the same for all sample sizes.

T F 6. By definition the degrees of freedom for all statistical tests equals $N - 1$.

T F 7. The mean of the *t* distribution equals 0 for all sample sizes.

T F 8. When df = ∞ the *t* distribution is identical to the *z* distribution.

T F 9. The *t* test can only be applied to a nondirectional alternative hypothesis.

T F 10. For a two-tailed test, if $|t_{obt}| \geq |t_{crit}|$, reject H_0.

T F 11. The *t* test requires that the sampling distribution of $\overline{X}$ is normal.

T F 12. The confidence interval is one example of point estimation.

T F 13. The 95% confidence interval means that 95% of the time $\overline{X} = \mu$.

T F 14. The 99% confidence limit is always wider than the 95% confidence interval.

T F 15. If r_{obt} is not equal to zero, ρ cannot equal zero.

T F 16. For the *t* test of the significance of the correlation coefficient, the statistic of interest is $\overline{X}$.

T F 17. If $r = -0.10$ and $N = 8$, ρ cannot equal +0.10.

T F 18. If we conclude that there is a significant correlation in the population we may be making a Type I error.

T F 19. For the distribution of $\overline{X}$ to be normally distributed, the distribution of X must be normally distributed.

T F 20. The larger the sample size the more likely $\overline{X}$ is close to μ.

T F 21. At $\alpha = 0.05$ there is more area in the critical region of rejection for the *t* distribution than for the *z* distribution.

T F 22. Changing sample size has no effect on power when using the *t* test.

T F 23. If $\rho = 1.00$, then r_{obt} must equal 1.00.

T F 24. Other variables held constant, the larger the size of the difference between $\overline{X}_{obt}$ and , the larger the size of effect.

T F 25 $\hat{d}$ measures both the size of effect and the direction of the effect.

T F 26 $\hat{d}$ gives a value in raw score units for the size of effect.

SELF-QUIZ

1. The 95% confidence interval of the mean for $\overline{X}$ = 13.0, s = 1.6, and N = 21 is ________.

   a. 12.01 to 13.99
   b. 12.27 to 13.73
   c. 11.05 to 15.05
   d. 12.95 to 14.95

2. How many t distributions are there?

   a. 1
   b. 2
   c. 30
   d. one for each value of df

3. The mean of the t distribution equals ________.

   a. 0
   b. 1

c. N
d. N – 1

4. For a given value of alpha, the critical value of t is _______ than the critical value of z.

a. no different
b. less
c. greater
d. none of the above

5. As N gets _______, the critical value of t gets _______.

a. larger; larger
b. smaller; smaller
c. larger; smaller
d. smaller; larger

6. As N gets infinitely large, the critical value of t equals the critical value of z.

a. true
b. false

7 If $\mu = 30$, □ = 5.2, $\overline{X} = 28.0$, $s = 6.1$ and $N = 13$, the value of the most powerful statistic to test the significance of the sample mean is _______.

a. –1.18
b. 1.96
c. 2.18
d. –1.39

8. If the population parameters are known, the t test is _______ powerful than the z test.

a. more
b. less
c. equally
d. need more information

9. As N increases $s_{\overline{X}}$ becomes a _______ estimate of $\sigma_{\overline{X}}$.

a. better
b. worse
c. meaningless
d. biased

10. The proper use of the t test requires that _______.

a. the sampling distribution of $\overline{X}$ is normal
b. $N \geq 30$
c. a and b
d. a or b

11. The _______ confidence interval for the population mean is always wider than the _______ confidence interval for the population mean.

a. 99%; 95%
b. 95%; 99%

12. The sample mean is an example of _______.

a. interval estimation
b. point estimation
c. confidence point
d. confidence numbers

13. For N = 20, the degrees of freedom for the test of the significance of the correlation coefficient (r) is _______.

    a. 20
    b. 21
    c. 19
    d. 18

14. For N = 20, the value of rcrit for □ = 0.052 tail is _______.

    a. 0.4329
    b. 0.3783
    c. 0.4438
    d. 0.5614

15. If r = 0.84 and N = 5, the value of tobt for the test of the significance of r is _______.

    a. 1.96
    b. 2.40
    c. 2.68
    d. 3.46

16. You conduct a single sample experiment to determine if the cingulate cortex is involved in learning tasks involving choice behavior. Twenty-two rats with lesions of the cingulate cortex are tested in a two choice Y-maze with the correct arm of the maze being randomly determined from trial-to-trial. A signal light illuminates the correct arm on each trial. Assume that cingulate cortex lesions do not interfere with reception of the signal light. Previous research with a large number of rats on this task has shown that the mean number of trials learn the task is 15 trials. The results of the experiment show a mean of 18 trials with a standard deviation of 5.8 for the lesioned rats to learn the task.

    a. What is your conclusion, using □ = 0.052 tail?
    b. What is the size of the effect?

ANSWERS

Exercises: 1a. 24 **1b.** 23 **1c.** 1 **1d.** 10 **1e.** 8 **2a.** –1.73 **2b.** ±2.201 **2c.** retain H_0 **3a.** ±2.201 **3b.** ±2.861 **3c.** ±12.706 **3d.** ±3.747 **3e.** ±2.878 **3f.** 2.179 **3g.** 2.921 **3h.** 1.895 **3i.** 2.624 **3j.** 1.833 **4a.** exercise reduces systolic blood pressure **4b.** exercise does not reduce systolic blood pressure **4c.** $t_{obt} = -2.74$ while $t_{crit} = -2.681$; therefore reject H_0 **4d.** $\hat{d} = 0.76$; medium effect. **5.** Since $t_{obt} = -4.61$ and $t_{crit} = \pm 2.776$, reject H_0, i.e. it is not reasonable **6a.** $t_{obt} = 1.42$ **6b.** retain H_0 since $t_{crit} = \pm 2.447$ **7a.** Reject H_0, $t_{obt} = -4.15$ and $t_{crit} = \pm 2.145$. Sleep affects memory. It appears to increase it. **7b.** $\hat{d} = 1.07$; large effect. **8a.** 102.92 to 106.58 **8b.** 102.04 to 107.46; using $s = 2.188$ **9a.** 46.67 to 53.33 **9b.** 45.42 to 54.58 **9c.** 45.21 to 54.79 **9d.** 43.11 to 56.89 **10a.** z test **10b.** 1.41 **10c.** retain H_0 ($z_{crit} = \pm 1.96$) **11a.** 1.25 **11b.** z generally does, but in this case neither does **11c.** Type II **12a.** $t_{obt} = 2.12$ **12b.** retain H_0 ($t_{crit} = \pm 2.145$) **12c.** 0.4973 **13a.** 0.895 **13b.** 0.8114 **13c.** yes.

True-False 1. F **2.** T **3.** T **4.** F **5.** F **6.** F **7.** T **8.** T **9.** F **10.** T **11.** F **12.** F **13.** F **14.** T **15.** F **16.** F **17.** F **18.** T **19.** F **20.** T **21.** F **22.** F **23.** T. **24.** T **25.** F **26.** F

Self-Quiz: 1. b **2.** d **3.** a **4.** c **5.** c **6.** a **7.** d **8.** b **9.** a **10.** d **11.** a **12.** b **13.** d **14.** c **15.** c **16a.** Reject H_0. $t_{obt} = 2.43$, $t_{crit} = \pm 2.080$. It appears that lesions of the cingulate cortex inhibit performance in this choice dependent task. **16b.** $\hat{d} = 0.52$; medium effect.

Chapter 14

STUDENT'S *t* TEST FOR CORRELATED AND INDEPENDENT GROUPS

CHAPTER OUTLINE

I. The Two-Condition Experiment

A. Types of two-condition experiments.

1. Correlated groups design. In the earlier chapters we analyzed this using the sign test.
2. Independent groups design. This design is covered later in the chapter.

B. Limitations of single sample design.

1. At least one population parameter (μ) must be specified.
2. Usually μ is not known.
3. Even if μ were known, one cannot be certain that the conditions under which μ was calculated are the same for a new set of experimental

4. These limitations are overcome in the two-condition experiment.

II. Student's *t* Test for Correlated Groups

A. Characteristics of repeated measures or correlated groups design.

1. Each subject used for both conditions (e.g. before and after; control and experimental).
2. Or pairs of subjects matched on one or more characteristics serve in both conditions.

B. Information used by correlated groups *t* test.

1. magnitude of difference scores.
2. direction of difference scores.

C. What is tested. Tests the assumption that the difference scores are a random sample from a population of difference scores having a mean of zero.

D. Similar to *t* test for single samples. The only change is that in this case we deal with difference scores instead of raw scores.

E. Equations.

$$t_{obt} = \frac{\overline{D}_{obt} - \mu_D}{s_D / \sqrt{N}}$$

$$t_{obt} = \frac{\overline{D}_{obt} - \mu_D}{\sqrt{\frac{SS_D}{N(N-1)}}}$$

where $D =$ difference score (e.g. control score - experimental score)

$\overline{D}_{\text{obt}} =$ mean of the sample difference scores

$\mu_{\text{D}} =$ mean of the population of difference scores (usually but not necessarily equal to 0)

$s_{\text{D}} =$ standard deviation of the sample difference scores

$SS_{\text{D}} =$ sum of squares of the sample difference scores

$= \Sigma D^2 - (\Sigma D)^2/N$

$N =$ number of difference scores

F. Size of Effect.

1. Rationale. As with the single sample *t* test, the statistic we are using to measure size of effect is symbolized by "*d*." It is a standardized statistic that with the correlated groups *t* test, relies on the relationship between the size of effect and $|\overline{D}_{\text{obt}}|$. As the size of effect gets greater, so does $|\overline{D}_{\text{obt}}|$, regardless of the direction of the effect. The statistic *d* uses the absolute value of $\overline{D}_{\text{obt}}$ since we are interested in the *size* of real effect, and are not concerned about direction. This allows *d* to have a positive value that increases with the size of $|\overline{D}_{\text{obt}}|$ regardless of the direction of the real effect. $|\overline{D}_{\text{obt}}|$ is divided by $_{\text{D}}$ to create a standardized value, much as was done with *z* scores.

2. Formula for Cohen's *d*.

$$d = \frac{|\overline{D}_{obt}|}{\sigma_D}$$ *Conceptual equation for size of effect, correlated groups t test*

Since $_D$ is unknown, we estimate it using s_D, the standard deviation of the sample difference scores. Substituting s_D for $_D$, we arrive at the computational equation for size of effect. Since s_D is an estimate, $\hat{d}$ is used instead of *d*.

$$\hat{d} = \frac{|\overline{D}_{obt}|}{s_D}$$ *Computational equation for size of effect, correlated groups t* test

3. Interpreting the value of $\hat{d}$. To interpret the value of $\hat{d}$, we are using the criteria that Cohen has provided. These criteria are given in the following table.

Value of $\hat{d}$	Interpretation of $\hat{d}$
0.00 – 0.20	Small effect
0.21 – 0.79	Medium effect
≥0.80	Large effect

G. Power. The correlated groups *t* test is more powerful than the sign test. Therefore there is less chance of making a Type II error. Note: As a general rule one uses the most powerful statistical analysis appropriate to the data.

H. Assumptions. For use requires that the sampling distribution of $\overline{D}$ is normally distributed. This can be achieved generally by

1. $N \geq 30$, or
2. Population scores normally distributed

III. Independent Groups Design

A. Design Characteristics

1. Random sampling of subjects from population.
2. Random assignment to each condition.
3. No basis for pairing of scores.
4. Each subject tested only once.
5. Raw scores are analyzed.
6. *t* test analyzes difference between sample means.

IV. Use of *z* Test for Independent Groups

A. Formula.

$$z_{obt} = \frac{(\overline{X}_1 - \overline{X}_2) - \mu_{\overline{X}_1 - \overline{X}_2}}{\sigma_{\overline{X}_1 - \overline{X}_2}}$$

where

$$\sigma_{\overline{X}_1 - \overline{X}_2} = \sqrt{\sigma_{\overline{X}_1}{}^2 + \sigma_{\overline{X}_2}{}^2} = \sqrt{\sigma^2\left(\frac{1}{n_1} + \frac{1}{n_2}\right)}$$

B. Assumptions.

1. Changing level of the independent variable is assumed to affect the mean of the distribution but not the standard deviation.
2. $\sigma_1{}^2 = \sigma_2{}^2 = \sigma^2$

C. Characteristics of sampling distribution of the difference between sample means.

1. Assuming population from which samples are drawn is normal, then the distribution of the difference between sample means is normal.
2. $\mu_{\overline{X}_1-\overline{X}_2} = \mu_1 - \mu_2$, where $\mu_{\overline{X}_1-\overline{X}_2}$ = the mean of the distribution of the difference between sample means
3. $\sigma_{\overline{X}_1-\overline{X}_2} = \sqrt{\sigma_{\overline{X}_1}{}^2 + \sigma_{\overline{X}_2}{}^2}$, where $\sigma_{\overline{X}_1-\overline{X}_2}$ = the standard deviation of the difference between sample means; $\sigma_{\overline{X}_1}$ = the variance of the sampling distribution of the mean for samples of size n_1 taken from the first population; and $\sigma_{\overline{X}_2}{}^2$ = the variance of the sampling distribution of the mean for samples of size n_2 taken from the second population.

D. Must Know σ^2. To use *z* test one must know σ^2 which is rarely the case.

V. Student's *t* Test for Independent Groups

A. Used when σ^2 must be estimated. Uses a weighted average of the sample variances, $s_1{}^2$ and $s_2{}^2$, as the estimate with df as the weights.

B. General equation.

$$t_{\text{obt}} = \frac{(\overline{X}_1 - \overline{X}_2) - \mu_{\overline{X}_1-\overline{X}_2}}{\sqrt{s_W{}^2\left(\frac{1}{n_1} + \frac{1}{n_2}\right)}} = \frac{\overline{X}_1 - \overline{X}_2}{\sqrt{\left(\frac{SS_1 + SS_2}{n_1 + n_2 - 2}\right)\left(\frac{1}{n_1} + \frac{1}{n_2}\right)}}$$

where $df = n_1 + n_2 - 2 = N - 2$

C. Equation when $n_1 = n_2$.

$$t_{obt} = \frac{\overline{X}_1 - \overline{X}_2}{\sqrt{\frac{SS_1 + SS_2}{n(n-1)}}}$$

D. Assumptions for use of *t* test for independent groups.

1. Sampling distribution of $\overline{X}_1 - \overline{X}_2$ is normally distributed, i.e. populations from which samples were taken must be normal.
2. $\sigma_1^2 = \sigma_2^2$ (homogeneity of variance).

E. Violations of assumptions. If $n_1 = n_2$ and $n \geq 30$, then the *t* test is robust if above assumptions are violated. If violations are extreme, use Mann-Whitney *U* test. This test is covered in Chapter 17.

VI. Size of Effect

A. Rationale. As with the correlated groups *t* test, the statistic we are using to measure size of effect is symbolized by "*d*." It is a standardized statistic that with the independent groups *t* test, relies on the relationship between the size of effect and $|\overline{X}_1 - \overline{X}_2|$. As the size of effect gets greater, so does $|\overline{X}_1 - \overline{X}_2|$, regardless of the direction of the effect. The statistic *d* uses the absolute value of $\overline{X}_1 - \overline{X}_2$ since we are interested in the *size* of real effect, and are not concerned about direction. This allows *d* to have a positive value that increases with the size of $|\overline{X}_1 - \overline{X}_2|$ regardless of the direction of the real effect. $|\overline{X}_1 - \overline{X}_2|$ is divided by to create a standardized value, much as was done with *z* scores.

B. Formula for Cohen's *d*:

$$\mathbf{d} = \frac{|\overline{\mathbf{X}}_1 - \overline{\mathbf{X}}_2|}{\sigma}$$ *Conceptual equation for size of effect, independent groups t test*

Since is unknown, we estimate it using $\sqrt{s_W^2}$, the weighted estimate of . Substituting $\sqrt{s_W^2}$ for , we arrive at the computational equation for size of effect. Since $\sqrt{s_W^2}$ is an estimate, $\hat{d}$ is used instead of *d*.

$$\hat{\mathbf{d}} = \frac{|\overline{\mathbf{X}}_1 - \overline{\mathbf{X}}_2|}{\sqrt{\mathbf{s_W}^2}}$$ *Computational equation for size of effect, independent groups t test*

C. Interpreting the value of $\hat{d}$. To interpret the value of $\hat{d}$, we again use the criteria that Cohen has provided. These criteria are shown below.

Value of $\hat{d}$	Interpretation of $\hat{d}$
0.00 – 0.20	Small effect
0.21 – 0.79	Medium effect
≥0.80	Large effect

VII. Power of *t* Test

A. Effect of variables on the power of the *t* test.

1. The greater the effect of the independent variable, the higher the power.
2. Increasing sample size increases power.
3. Increasing sample variability decreases power.

VIII. Use of Correlated or Independent *t*

A. Which test to use.

1 Correlated *t* advantageous when there is a high correlation between the paired scores.
2 Correlated *t* is advantageous when there is low variability in difference scores and high variability in raw scores.
3. Independent *t* is more efficient from df per measurement analysis.
4. Some experiments do not allow same subject to be used twice (i.e. comparing males vs. females), so must use independent *t* test.

IX. Alternative Approach using Confidence Intervals

A. Null Hypothesis Approach. Evaluate the probability of getting obtained results or results even more extreme assuming chance alone is operating. If obtained probability $\leq$, reject H_0.

B. Confidence Interval Approach. Uses confidence intervals to determine if it is reasonable to reject H_0 and at the same time gives an estimate of the size of the real effect.

C. <u>95% Confidence Interval for $_1 - _2$</u>. By estimating the 95% confidence interval for $_1 - _2$, we can determine if it is reasonable to reject H_0 for an alpha level = 0.05 and if so, the confidence interval can be used as an estimate of the size of the real effect. We have 95% confidence that the interval $_1 - _2$ contains the size of the real effect.

D. <u>Equations for constructing the 95% Confidence Interval for $_1 - _2$</u>.

$$\mu_{\text{lower}} = (\overline{X}_1 - \overline{X}_2) - s_{\bar{X}_1 - \bar{X}_2} t_{0.025}$$
$$\mu_{\text{upper}} = (\overline{X}_1 - \overline{X}_2) + s_{\bar{X}_1 - \bar{X}_2} t_{0.025}$$

where

$$s_{\bar{X}_1 - \bar{X}_2} = \sqrt{\left(\frac{SS_1 + SS_2}{n_1 + n_2 - 2}\right)\left(\frac{1}{n_1} + \frac{1}{n_2}\right)}$$

E. <u>If the interval $_1 - _2$ contains the value "0"</u>. If the interval $_1 - _2$ contains the value "0", we cannot reject H_0 at = 0.05.

F. <u>99% confidence interval for $_1 - _2$</u>. By estimating the 99% confidence interval for $_1 - _2$, we can determine if it is reasonable to reject H_0 for an alpha level = 0.01 and if so, the confidence interval can be used as an estimate of the size of the real effect. In this case, we have 99% confidence that the interval $_1 - _2$ contains the size of the real effect.

G. <u>Equations for constructing the 99% Confidence Interval for $_1 - _2$</u>.

$$\mu_{\text{lower}} = (\overline{X}_1 - \overline{X}_2) - s_{\overline{X}_1 - \overline{X}_2}\, t_{0.005}$$

$$\mu_{\text{upper}} = (\overline{X}_1 - \overline{X}_2) + s_{\overline{X}_1 - \overline{X}_2}\, t_{0.005}$$

H. <u>If the interval $_1 - _2$ contains the value "0"</u>. If the interval $_1 - _2$ contains the value "0", we cannot reject H_0 at $= 0.01$.

CONCEPT REVIEW

In this chapter we have been studying

methods of analyzing

data from the (1) _______ condition experiment. (1) two

It has great advantage over the one condition

experiment. We do not need to specify

any (2) _______ parameters to analyze (2) population

the results.

In the repeated measures or (3) _______ (3) correlated

(4) _______ design, a (5) _______ (4) groups (5) difference

score is calculated and analyzed. Though it is

common, it is not necessary that the	
(6) _______ subject be used under both	(6) same
conditions. We can (7) _______ the	(7) match
subjects on certain important characteristics	
instead. The (8) _______ test utilizes	(8) sign
information only on the (9) _______ of the	(9) direction
difference between scores; whereas the *t* test	
for correlated groups utilizes information on	
both direction and (10) _______ of the	(10) magnitude
difference scores. In evaluating H_0, this *t* test	
usually tests the assumption that the difference	
scores are a random (11) _______ from a	(11) sample
population of difference scores having a	
mean of (12) _______. The equation for the *t*	(12) zero
test for correlated groups is:	

$$t_{\text{obt}} = \frac{(13)\_\_\_\_\_ - \mu_D}{\frac{(14)\_\_\_\_\_}{(15)\_\_\_\_\_}}$$

(13) $\overline{D}$
(14) s_D
(15) N

The computational formula is:

$$t_{obt} = \frac{(16)\_\_\_\_\_ - \mu_D}{\sqrt{\frac{(17)\_\_\_\_\_}{(18)\_\_\_\_\_(N-1)}}}$$

(16) $\overline{D}$
(17) SS_D
(18) N

where

(19) _______ = the difference score, (19) D

(20) _______ = the mean of the sample difference scores, (20) $\overline{D}$

(21) _______ = mean of the population of difference scores, (21) μ_D

(22) _______ = the standard deviation of the sample difference scores, (22) s_D

(23) _______ = number of difference scores, (23) N

(24) _______ = $D - \overline{D}$ (24) d

For all t tests the decision rule for evaluating the null hypothesis is the same:

If | (25) _______ | ≥ | (26) _______ |,

(25) t_{obt}
(26) t_{crit}

(27) _______ H_0 (27) reject

Size of Effect Using Cohen's *d*

For the correlated groups *t* test,

Cohen's *d* uses the absolute

value of (28) ________ to measure the size — (28) $\overline{D}_{\text{obt}}$

of effect, because the size of effect varies

(29) ________ with this absolute value. — (29) directly

Rather than using raw score units, *d* is

a (30) ________ value — (30) standardized

achieved by dividing $\overline{D}_{\text{obt}}$ by (31) ________. — (31) $_{\text{D}}$

The conceptual formula for *d* is given by

$$d = \frac{|(32)\ \_\_\_\_\_\_|}{(33)\ \_\_\_\_\_\_}$$

(32) $\overline{D}_{\text{obt}}$

(33) $_{\text{D}}$

Since $_{\text{D}}$ is unknown, we estimate it

with (34) ________, resulting — (34) s_{D}

in the following computational equation

for size of effect.

$$\hat{d} = \frac{|(35)\ \_\_\_\_\_\_|}{(36)\ \_\_\_\_\_\_}$$

(35) $\overline{D}_{\text{obt}}$

(36) s_{D}

Compared to the sign test, the *t* test	
is (37) _______ powerful.	(37) more
There is a (38) _______ probability of	(38) lower
making a Type II error. As a general rule,	
one should use the most (39) _______	(39) powerful
test available to analyze the results of an	
experiment since this gives the highest	
(40) _______ of rejecting H_0. To use the *t* test	(40) prob-ability
for correlated groups, the (41) _______	(41) sampling
(42) _______ of *D* must be (43) _______	(42) distribution
	(43) normally
distributed. This means *N* should be ≥	
(44) _______, or that the population scores	(44) 30
themselves should	
be normally distributed.	
The other major type of experimental	
design is called the (45) _______	(45) independ ent

(46) _______ design. Here each subject is -	(46) groups
tested only (47) _______. This design has	(47) once
(48) _______ or more groups, usually an	(48) two
(49) _______ group and a	(49) experi-mental
(50) _______group. Subjects are (51) _______	(50) control (51) randomly
selected from the (52) _______ and then	(52) population
randomly divided into the two (or sometimes	
more) groups. There is (53) _______ basis for	(53) no
pairing scores between the two conditions.	
A comparison is made between the scores	
of each (54) _______ to determine if	(54) group
chance alone is a reasonable explanation	
of the (55) _______ between the group	(55) differences
scores. The scores are analyzed as separate	
(56) _______. If chance alone is the correct	(56) samples
explanation of the results then sets of sample	
scores (groups) can be considered	
random samples from the (57) _______	(57) same or

identical

populations.

With the *z* test for independent groups, we analyze the (58) _______ between the sample (59) _______. In order to use the *z* test to analyze the data from an independent groups design, we would need to know the value of the population parameter (60) _______. Since it is rarely known, we estimate it using the weighted (61) _______ of the sample variances. The weighting is done using the (62) _______ (63) _______ as the weights. The formula for the *t* test for independent groups is:

$$t_{\text{obt}} = \frac{((64)\_\_\_\_ - (65)\_\_\_\_) - \mu_{\overline{X}_1 - \overline{X}_2}}{\sqrt{(66)\_\_\_\_\left(\frac{1}{(67)\_\_\_\_} + \frac{1}{(68)\_\_\_\_}\right)}}$$

(58) difference
(59) means
(60) σ^2
(61) average
(62) degrees
(63) of freedom
(64) $\overline{X}_1$
(65) $\overline{X}_2$
(66) $s_W^{\ 2}$
(67) n_1
(68) n_2

Generally, we test the null hypothesis by

assuming that $\mu_{\overline{X}_1-\overline{X}_2}$ equals (69) ________. (69) 0

Also $s_W{}^2$ is the (70) ________ (70) weighted

estimate of (71) ________. The general (71) σ^2

computational formula is:

$$t_{\text{obt}} = \frac{\left(\overline{X}_1 - \overline{X}_2\right) - \mu_{\overline{X}_1-\overline{X}_2}}{\sqrt{\left(\frac{(72)\_\_ + (73)\_\_}{(74)\_\_ + (75)\_\_ - (76)\_\_}\right)\left(\frac{1}{(77)\_\_}\right)}}$$

(72) SS_1
(73) SS_2
(74) n_1
(75) n_2
(76) 2
(77) n_1
(78) n_2

where $SS =$ (79) ___ – (80) ___/ (79) ΣX^2
(80) $(\Sigma X)^2$

(81) ___ for each sample. (81) (N)

The df = N – (82) ________ where (82) 2

$N =$ (83) ________ + (83) n_1

(84) ________. (84) n_2

The computational formula when $n_1 = n_2$ is:

$$t_{\text{obt}} = \frac{(\overline{X}_1 - \overline{X}_2)}{\sqrt{\dfrac{SS_1 + SS_2}{(85)\_\_\_\_ ((86)\_\_\_\_ - (87)\_\_\_\_)}}}$$

(85) *n*
(86) *n*
(87) 1

If n_1 (88) _______ n_2, the more general equation must be used. (88) $\neq$

To use the *t* test for independent groups we assume that the sampling distribution of (89) _______ is normally distributed. (89) $\overline{X}_1 - \overline{X}_2$

We also assume (90) _______ of variance. (90) homogeneity

One way for this to occur is if the (91) _______ variable affects the means of the populations but not their (92) _______ (93) _______. The *t* test is considered to be (94) _______, meaning it withstands violations of the above assumptions under some conditions. The *t* test is relatively (95) _______ to violations of the assumptions if

(91) independent
(92) standard
(93) deviations
(94) robust
(95) insensitive

(96) _______ = (97) _______ and each | (96) n_1
(97) n_2

sample ≥ (98) _______. | (98) 30

In general with the *t* test, power can be affected by several factors. The (99) _______ the effect of the independent variable, the more likely the value of the (100) _______ of the *t* equation will be large. If other factors are constant this will increase the power.

(99) greater

(100) numerator

Increasing the sample size will (101) _______ the denominator thereby (102) _______ power.

(101) decrease

(102) increase-ing

As the sample variance increases the denominator of the *t* equation will (103) _______ which will (104) _______ power.

(103) increase

(104) decrease

Size of Effect Using Cohen's *d*

For the independent groups *t* test, Cohen's *d* uses the absolute value of (105) ________ to measure the size — (105) $\overline{X}_1 - \overline{X}_2$

of effect, because the size of effect varies

(106) ________ with this absolute value. — (106) directly

Rather than using raw score units, *d* is a

(107) ________ value — (107) standardized

achieved by dividing $\overline{X}_1 - \overline{X}_2$

by (108) ________. — (108)

The conceptual formula for *d* is given by

$$d = \frac{|(109)\_\_\_\_\_ - (110)\_\_\_\_\_|}{(111)\_\_\_\_\_}$$

(109) $\overline{X}_1$
(110) $\overline{X}_2$
(111)

Since   is unknown, we estimate it with

(112) ________, — (112) $\sqrt{s_W^2}$

resulting in the following computational equation for size of effect.

$$\hat{d} = \frac{|(113)\_\_\_\_\_ - (114)\_\_\_\_\_|}{(115)\_\_\_\_\_}$$

(113) $\overline{X}_1$
(114) $\overline{X}_2$
(115) $\sqrt{s_W^2}$

There are certain advantages and

disadvantages to using

the *t* test for correlated or independent	
groups. When there is (116) _______	(116) high
variability in the (117) _______ scores, but	(117) raw
(118) _______ variability in the	(118) low
(119) _______ scores, it would	(119) difference
be preferable to use the *t* test for correlated	
groups. If the (120) _______	(120) correlation
(121) _______ between the paired scores of	
condition 1 and 2 is high it would be advantageous	
to use the *t* test for correlated groups.	(121) coefficient
Many times it is not possible to use	
the same (122) _______ in both conditions	(122) subject
and often it is difficult to achieve appropriate	
matching. Under these conditions we would	
use the *t* test for (123) _______ (124) _______	(123) independent (124) groups
The *t* test for independent groups has as its	
major advantages the fact that it makes more	
efficient use of (125) _______ of	(125) degrees

(126) _______ for each measurement and (126) freedom

its relative ease of recruiting (127) _______. (127) subjects

Confidence Interval Approach for Evaluating the Effect of the Independent Variable in the Two Group, Independent Groups Design

This approach uses (128) _______ (128) confidence

(129) _______ to determine if we should (129) intervals

reject H_0 and also gives an estimate of the

(130) _______ of the real effect. Following (130) size

this approach, we

estimate the 95% or 99% confidence

interval for (131) _______. (131) $\mu_1 - \mu_2$

If the 95% confidence interval contains

the value (132) ______, (132) 0

we cannot reject H_0 at = (133) _______. (133) 0.05

If the 99% confidence interval contains the

value (134) _______, we cannot (134) 0

reject H_0 at = (135) _______. If the (135) 0.01

confidence interval doesn't contain the

value "0", than we (136) _______ reject H_0, (136) can

and the interval gives us an estimate

of the (137) _______ of (137) size

the real effect. The real effect lies somewhere

in the interval, always keeping in mind the

possibility of a Type 1 error. The equations for

constructing the 95% confidence interval for

$\mu_1 - \mu_2$ are

μ_{lower} = (138) (_______) – (139) _______ x (140) _______

(138) $\bar{X}_1 - \bar{X}_2$

(139) $s_{\bar{X}_1 - \bar{X}_2}$

(140) $t_{0.025}$

μ_{upper} = (141) (_______) + (142) _______ x (143) _______

(141) $\bar{X}_1 - \bar{X}_2$

(142) $s_{\bar{X}_1 - \bar{X}_2}$

(143) $t_{0.025}$

where

$$s_{\bar{X}_1 - \bar{X}_2} = \sqrt{\left(\frac{(144)\_\_ + (145\_\_}{(146)\_\_ + (147)\_\_ - 2}\right)\left(\frac{1}{(148)\_\_} + \right.}$$

(144) SS_1

(145) SS_2

(146) n_1

(147) n_2

(148) n_1

(149) n_2

The equations for constructing the 99%

confidence interval for

$\mu_1 - \mu_2$ are

μ_{lower} = (150) (________) – (151) ________ x (152) ________

(150) $\bar{X}_1 - \bar{X}_2$

(151) $s_{\bar{X}_1 - \bar{X}_2}$

(152) $t_{0.005}$

μ_{upper} = (153) (________) + (154) ________ x (155) ________

(153) $\bar{X}_1 - \bar{X}_2$

(154) $s_{\bar{X}_1 - \bar{X}_2}$

(155) $t_{0.005}$

EXERCISES

1. A dog food manufacturer believes that he may have found a way to improve the taste of the dog food made by his company. To test the new process, the company recruited 12 dogs. The dogs were given the original food and the time required to eat a standard portion of the food was measured. Two days later the same dogs were fed the same quantity of the new formula and the time was measured again. The following times were recorded:

Dog	Old Process (sec)	New Process (sec)
A	135	120
B	115	118
C	120	120
D	140	120
E	137	131
F	150	142
G	124	125
H	132	121
I	119	117
J	147	148
K	160	135
L	135	128

It is assumed that faster eating represents greater preference for the product.

a. State the directional alternative hypothesis.
b. State the null hypothesis.
c. What is the value of t_{obt}?
d. What do you conclude using $\alpha = 0.05_{1\ \text{tail}}$?
e. If there is a real effect, compute the size of the effect, using $\hat{d}$. Is it a small, medium or large effect?
f. What is one obvious (okay, maybe it's not so obvious) defect with this study?

2. A political candidate wishes to determine if endorsing increased social spending is likely to affect her standing in the polls. She has access to data on the popularity of several other candidates who have endorsed increased spending. The data was available both before and after the candidates announced their positions on the issue. The data are as follows:

Candidate	Popularity Ratings Before	After
1	42	43
2	41	45
3	50	56
4	52	54
5	58	65
6	32	29
7	39	46
8	42	48
9	48	47
10	47	53

Assuming no other factors were influencing the popularity ratings;

a. State the nondirectional alternative hypothesis.
b. State the null hypothesis.
c. What is the value of t_{obt}?
d. What might the candidate conclude using $\alpha = 0.01_{2\ tail}$?
e. What type error might the candidate have made?

3. A medical researcher wishes to evaluate the effectiveness of a drug in prolonging the life of a red cell in a culture medium. Seven cultures are treated with saline (an inactive control substance) and seven others are treated with the new drug. The following survival times in days were recorded:

Control Group	Treatment Group
109	121
112	119
118	119
116	126
121	121
100	115
113	117

a. What kind of design is this?
b. State the directional alternative hypothesis.
c. State the null hypothesis.
d. What is the value of t_{obt}?
e. What do you conclude using $\alpha = 0.05_{1\ tail}$?
f. If there is a real effect, compute the size of the effect using $\hat{d}$? Is it a small, medium, or large effect.

4. A worker in a neighborhood clinic wishes to assess the impact of showing an educational film on patient compliance in taking an antihypertension medication. The diastolic blood pressure (BP) is used as the dependent variable. During the study the medication dosage was kept constant. Blood pressure was measured one week prior to the film, then the film was shown, and the blood pressure measured again 3 weeks later. The data are shown below:

	Diastolic BP (mmHg)	
Patient	**Before**	**After**
1	110	100
2	105	95
3	98	88
4	100	92
5	89	83
6	82	86
7	113	100
8	102	101
9	101	96
10	118	112

a. State the nondirectional alternative hypothesis.
b. State the null hypothesis.
c. What is the value of t_{obt}?
d. What do you conclude using $\alpha = 0.01_{2\ tail}$?

5. Consider the following data collected from a repeated measures design study.

Before	After
5	17
8	25
5	19
7	32
6	26
9	8
10	9

a. Using $\alpha = 0.05_{2\ tail}$, what would you conclude about the null hypothesis using the *t* test?
b. What would you conclude using the sign test?

c. What does this indicate?

6. A biologist believes that temperature affects the croaking (noise making, not dying) behavior of frogs. A group of laboratory frogs are randomly divided into 2 groups and placed in identical terrariums. The control group of frogs' is kept at a constant 22°C. The experimental group is kept at a temperature of 30° C. The number of croaks emitted over a 10-minute period are counted. The data are as follows:

Number of Croaks	
22 °	**30 °**
23	**30**
30	**32**
31	**36**
28	**39**
26	**30**
12	**18**
19	**25**
18	**26**

a. State the nondirectional alternative hypothesis.
b. State the null hypothesis.
c. What is the value of t_{obt}?
d. What do you conclude using $\alpha = 0.01_{2\ tail}$?
e. Can you think of another way to design this study to get more information for the biologist? Explain.

7. A teacher wishes to test the effects of using reinforcement versus a conventional method for maintaining classroom deportment. Twenty different teachers are instructed on the techniques to be used in the study; 10 on the conventional method and 10 on using reinforcement. Data was gathered on the number of minutes of attentive behavior in an hour sample taken from a class of each teacher when the classes

were being taught the same material. The classrooms were chosen because they were not significantly different in terms of socioeconomic status or past educational experiences. The data are as follows:

Attentive Behavior (Min)	
Conventional	**Reinforcement**
12	30
20	41
25	32
14	15
16	47
40	50
38	39
36	35
29	20
33	40

a. State the nondirectional alternative hypothesis.
b. State the null hypothesis.
c. What is the value of t_{obt}?
d. What do you conclude using $\alpha = 0.01_{2\ tail}$?

8. As an owner of a computer component factory you want to buy a new machine to make integrated circuits. You can buy one of two types of machines. You obviously want to buy the better machine (we'll assume they are comparable in price and reliability). You rent both machines for 8 days and make sample runs on both machines. The following data are the number of defective components per 1000 on each day.

Machine A	Machine B
2	5
1	3
3	5
1	4
2	4
4	7
6	9
0	4

a. What is the value of t_{obt}?

b. Which machine would you buy? Use $\alpha = 0.05_{2\text{ tail}}$. Assume independence between each day's run of each machine.

c. Determine the size of the real effect, using $\hat{d}$.

d. Using Cohen's criteria for size of effect, is this a small, medium or large effect?

e. Construct the 95% confidence interval for ${}_1 - {}_2$. Using this interval, what do you conclude regarding H_0? Using the confidence interval approach, what is your conclusion about the size of effect?

9. A drug company wants to evaluate the effects of a new tonic on stimulating new hair growth. Two comparable groups of male volunteers are tested. Group 1 is given a mixture of chicken fat and bear grease as a control tonic. Group 2 gets the new tonic. The number of new hairs are counted one month after the onset of tonic use. The results are shown below:

New Tonic	Control Tonic
6	5
12	13
10	6
14	8
13	25
8	50
65	

a. State the directional alternative hypothesis.
b. State the null hypothesis.
c. What is the value of t_{obt}?
d. What do you conclude? $\alpha = 0.05_{1\ tail}$.
e. What type error might you be making?

10. An automotive engineer wishes to determine the fuel efficiency of a test engine compared to the standard production model. The following data (miles/gallon) were obtained under standard conditions. Assume independence between scores in each group.

Engine	
Standard	**Test**
25	26
24	27
19	20
21	22
22	23
26	27
25	26
23	

a. Test the null hypothesis that there is no difference in the fuel efficiencies of these two engines. What do you conclude using $\alpha = 0.05_{2\ \text{tail}}$?
b. What type of error might you be making?

TRUE-FALSE QUESTIONS

T F 1. In a repeated measures design the difference scores are used in the analysis, not the raw scores.

T F 2. If H_0 is true, then μ_D must equal 0 for a correlated groups design.

T F 3. H_1 cannot be true if $\overline{D}_{obt} = 0$ for a correlated groups design.

T F 4. $SS_D = \sum(D - \overline{D})^2$

T F 5. When using the correlated groups *t* test, the larger the size of effect the larger is $\overline{D}_{obt}$

T F 6 Used in conjunction with the correlated groups *t* test, $\hat{d}$ only measures the direction of the effect.

T F 7 Used in conjunction with the correlated groups *t* test, $\hat{d}$ gives a value in standard score units for the size of effect.

T F 8. The *t* test for correlated groups is generally more powerful than the sign test.

T F 9. When several tests are appropriate for analyzing data, it is best to use the least powerful test so as to minimize the probability of making a Type I error.

T F 10. The *t* test for correlated groups requires the sampling distribution of $\overline{X}$ be normally distributed.

T F 11. In the independent groups design there is no matching of subjects and each subject is tested only once.

T F 12. With the *t* test for correlated groups, the difference between the sample means always equals the mean of the difference scores.

T F 13. One way of stating H_0 for the *t* test for independent groups is $\mu_1 - \mu_2 = 0$. (Assume H_1 is nondirectional)

T F 14. In the independent groups *t* test, $s_{\overline{X}_1-\overline{X}_2}$ is an estimate of $\sigma_{\overline{X}_1-\overline{X}_2}$.

T F 15. In the independent groups *t* test an average of s_1^2 and s_2^2, weighted by the degrees of freedom, is used to estimate σ^2.

T F 16. For both the *t* test for repeated measures and independent groups, df = $n_1 + n_2 - 2$.

T F 17. If $n_1 = n_2$ and σ^2 is unknown, then the *z* test can be used to analyze the data from an independent groups design.

T F 18. The *t* test for independent groups assumes homogeneity of variance.

T F 19. The *t* test assumes that the independent variable affects either the mean of the dependent variable or the standard deviation, but not both.

T F 20. If a violation of the homogeneity of variance assumption occurs, it is always necessary to abandon use of the *t* test for an independent groups design.

T F 21. Other factors held constant, for an independent groups design, the larger the value of $\overline{X}_1 - \overline{X}_2$, the greater the likelihood of rejecting H_0.

T F 22. For the independent groups design, increasing sample variance decreases power.

T F 23. Generally, the higher the magnitude of *r* between the scores of both groups the more powerful is the *t* test for correlated groups.

T F 24. If *r* between the scores of both groups is high and negative then the *t* test for correlated samples is not a powerful statistical test.

T F 25. The higher df, the lower t_{crit} becomes for rejecting H_0.

T F 26. There is a greater chance of making a Type I error using $\alpha = 0.05$ when using the *t* test for correlated groups then when using the sign test on the same scores.

T F 27. It is not necessary to measure the population parameters to apply the *t* test for either the independent or correlated groups designs.

T F 28. If $\overline{X}_1 - \overline{X}_2 = 0$, then H_0 must be true.

T F 29. When a result is statistically significant, that guarantees that the result is an important one.

T F 30. Assuming the variability of the scores doesn't change, when using the independent groups *t* test, the larger the size of effect the larger is $|\overline{X}_1 - \overline{X}_2|$.

T F 31 $\hat{d}$ can be used to measure the size of effect in conjunction with the single samples *t* test, the correlated groups *t* test, and the independent groups *t* test.

T F 32 For the independent groups *t* test, $\hat{d}$ gives a value in raw score units for the size of effect.

T F 33. When using the confidence interval approach for evaluating the effect of the independent variable, if the 95% confidence interval contains the value "0", we can reject H_0 at = 0.05.

T F 34. The confidence interval approach for evaluating the effect of the independent variable does not allow evaluation of the size of real effect.

SELF-QUIZ

1. A major advantage to using a two condition experiment (e.g. control and experimental groups) is _______.

   a. the test has more power
   b. the data are easier to analyze
   c. the experiment does not need to know population parameters
   d. a and b

2. In testing the null hypothesis, the correlated *t* test allows one to utilize information on _______ in the test of significance.

   a. magnitude and direction
   b. magnitude only
   c. direction only
   d. separation between the groups

3. In a correlated *t* test, if the independent variable has no effect, the sample difference scores are a random sample from a population where the mean difference score (μ_D) equals _______.

   a. 0
   b. 1
   c. *N*
   d. cannot be determined

4. In analyzing the results of the correlated *t* test, the _______ scores are analyzed.

   a. standardized
   b. normalized
   c. raw
   d. difference

5. The *t* test is _______ likely to result in a Type II error than the sign test for a repeated measures design.

   a. more
   b. less
   c. equally
   d. probably

6. Which of the following tests analyzes the difference between the means of two independent samples?

a. correlated *t* test
b. *t* test for independent groups
c. sign test
d. all of the above

7. In an independent groups design the nondirectional alternative hypothesis states _______.

a. $\mu_1 > \mu_2$
b. $\mu_1 < \mu_2$
c. $\mu_1 \neq \mu_2$
d. $\mu_1 = \mu_2$

8. The *z* test for independent groups is almost never used because it requires that _______ be known.

a. μ_1
b. $\overline{X}_1$
c. σ^2
d. s^2

9. In the *t* test for independent samples, there are _______ degrees of freedom.

a. $n_1 - 1$
b. $n_1 + n_2$
c. $n_1 - n_2 + 2$
d. $n_1 + n_2 - 2$

10. The *t* test assumes that the independent variable affects the _______ of the populations.

a. means

b. standard deviations
c. a and b
d. none of the above

11. The use of the *t* test for independent groups assumes _______.

a. $\overline{X}_1 - \overline{X}_2$ is normally distributed
b. $\sigma_1^2 = \sigma_2^2$
c. $\overline{D}$ is normally distributed
d. all of the above
e. a and b

12. The power of a correlated *t* test increases if the correlation between the paired scores is _______.

a. 0
b. high
c. low
d. cannot be determined

13. The sampling distribution of $\overline{X}_1 - \overline{X}_2$ has a mean value equal to _______.

a. 0
b. $\mu_1 - \mu_2$
c. N
d. $s_{\overline{X}_1 - \overline{X}_2}$

14. If $n_1 = n_2$ and n is relatively large, then the *t* test is relatively robust against _______.

a. violations of the assumptions of homogeneity of variance and normality
b. violations of random samples
c. traffic violations
d. violations by the forces of evil

15. Five students were tested before and after taking a class to improve their study habits. They were given articles to read which contained a known number of facts in each story. After the story each student listed as many facts as he/she could recall. The following data was recorded.

Before	**10**	**12**	**14**	**16**	**12**
After	**15**	**14**	**17**	**17**	**20**

What do you conclude using $\alpha = 0.05_{2\ \text{tail}}$?

a. reject H_0
b. retain H_0

16. The size of effect for the improvement class in problem **15**, using $\hat{d}$ equals _______.

a. 0.56
b. 0.52
c. 1.42
d. 1.37

17. Using Cohen's criteria, the size of effect found in **#16** is _______.

a. small
b. medium
c. large

d. Cohen's criteria doesn't apply

18. A physical therapist wants to know if football players (guards in this experiment) recover full strength following injuries. A group of injury free guards were given a strength test as were a group of guards who had finished a rehabilitation program following an injury. The groups were matched in height and weight. The following strength ratings were recorded.

Non-injured	Injured
302	297
306	300
258	250
239	245
280	270
265	262
274	

What do you conclude using $\alpha = 0.01_{2\ \text{tail}}$?

a. reject H_0
b. retain H_0

19. In an independent groups experiment, the size of the effect of the independent variable is estimated using _______.

a. t_{obt}
b. r_{obt}
c. $\hat{d}$
d. r^2

For questions 20 - 23, use the following experiment.

A school psychologist is interested in determining if children with attention deficit disorder (ADD) learn better if English literature is read to them rather than having them read the material themselves. A random sample of 10 sixth graders with ADD is selected and divided into two groups of 5 each. One of the groups has a story read to them (Listening Group) and the other reads the story themselves (Reading Group). A quiz on the story is given after each group has finished reading or hearing the story. The following scores were obtained with 20 being a perfect score.

Reading Group	Listening Group
8	15
10	12
7	13
12	11
6	10

20. What do you conclude, using $\alpha = 0.05_{\text{2 tail}}$?

    a. Reject H_0
    b. Retain H_0

21. The size of effect, using $\hat{d}$ equals _______.

    a. 1.72
    b. 0.58
    c. 0.32
    d. 1.65

22. Using Cohen's criteria, the size of effect found in **#21** is _______.

    a. small
    b. medium
    c. large

d. Cohen's criteria doesn't apply

23. The 95% confidence interval for $\mu_1 - \mu_2$ equals _______.

    a. 0.25 – 6.72
    b. 0.31 – 4.47
    c. 0.42 – 6.78
    d. 0.47 – 5.64

ANSWERS

Exercises: 1a. The change in manufacturing process reduces the time it takes dogs to eat a meal. $\mu_D > 0$ **1b.** The change in manufacturing process has no effect on the time dogs take to eat a meal. $\mu_D \leq 0$ **1c.** $t_{obt} = 2.88$ **1d.** $t_{crit} = 1.796$; therefore reject H_0 1e. $\hat{d} = 0.83$, large effect. **1f.** The order of food presentation is always the same. It would be better if half the dogs were given the new process first so if conditions in the lab are different on different days, it won't have a systematic effect on the results.

2a. Endorsing increased social spending affects popularity ratings. $\mu_D \neq 0$ **2b.** Endorsing increased social spending has no effect on popularity ratings. $\mu_D = 0$ **2c.** -3.10 **2d.** $t_{crit} = \pm 3.250$; retain H_0 **2e.** Type II

3a. Independent groups design **3b.** The new drug prolongs the life of red cells in a culture medium. $\mu_1 > \mu_2$ **3c.** The new drug does not prolong the life of red cells in a culture medium $\mu_1 \leq u_2$ **3d.** $t_{obt} = -2.40$ **3e.** $t_{crit} = -1.782$; reject H_0 **3f.** $\hat{d} = 1.28$, large effect.

4a. The film affects patient compliance. $\mu_D \neq 0$ **4b.** The film has no effect on patient compliance as measured by diastolic blood pressure; any change was due to chance alone. $\mu_D = 0$. **4c.** $t_{obt} = 4.12$ **4d.** $t_{crit} = \pm 3.250$; reject H_0

5a. $t_{obt} = -3.25$, $t_{crit} = \pm 2.447$; reject H_0 **5b.** $0.4532 > 0.05$; retain H_0 **5c.** *t* test for correlated measures has more power.

6a. Temperature affects the croaking behavior of laboratory frogs. **6b.** Temperature has no effect on the croaking behavior of laboratory frogs. $\mu_1 = \mu_2$ **6c.** $t_{obt} = -1.85$ **6d.** $t_{crit} = \pm 2.977$; retain H_0 **6e.** Use a regression experiment. Take the same frogs and measure the number of croaks at several different temperatures. Construct the regression line through the points (X = temp., Y = # of croaks). One could also calculate the correlation coefficient and test the significance of r.

7a. There is a difference in the attentive behavior of the two classes. **7b.** There is no difference in the attentive behavior of the two classes. $\mu_1 = \mu_2$. NOTE: The experiment is making the supposition that any difference is due to the use of either reinforcement or conventional techniques and not any other reason. This may not be a good assumption. **7c.** $t_{obt} = -1.79$ **7d.** $t_{crit} = \pm 2.878$; retain H_0

8a. $t_{obt} = -2.83$; $t_{crit} = \pm 2.145$ **8b.** Reject H_0. Machine A produces significantly fewer defective components than Machine B. I would buy Machine A. **8c.** $\hat{d} = 1.42$ **8d.** Using Cohen's criteria, this is a large effect. **8e.** The 95% confidence interval for ${}_1 - {}_2 = -4.83$ - -0.67; As in 8b, I would reject H_0. I am 95% confident that the interval of -4.83 - -0.67 contains the real effect of the difference between Machine A and Machine B. If so, than the real effect is that Machine A produces between 0.67 and 4.83 less defects/1000 than Machine B.

9a. The new hair tonic increases new hair growth. $\mu_1 > \mu_2$ **9b.** The new tonic does not increase new hair growth. $\mu_1 \leq \mu_2$ **9c.** $t_{obt} = -.04$ **9d.** $t_{crit} = 1.796$; retain H_0 **9e.** Type II.

10a. $t_{obt} = -0.99$: $t_{crit} = 2.160$. Retain H_0. **10b.** Type II.

True-False: **1.** T **2.** F **3.** F **4.** T **5.** T **6.** F **7.** T **8.** T **9.** F **10.** F **11.** T **12.** T **13.** T **14.** T **15.** T **16.** F **17.** F **18.** T **19.** F **20.** F **21.** T **22.** T **23.** T **24.** F **25.** T **26.** F **27.** T **28.** F **29.** F **30.** T **31.** T **32.** F **33.** F **34.** F.

Self-Quiz: **1.** c **2.** a **3.** a **4.** d **5.** b **6.** b **7.** c **8.** c **9.** d **10.** a **11.** e **12.** b **13.** b **14.** a **15.** a (t_{obt}= -3.062; $t_{crit} = \pm 2.776$) **16.** d **17.** c **18.** b (t_{obt} = 0.319; $t_{crit} = \pm 3.106$) **19.** c. **20.** a ($t_{obt} = 2.612$; $t_{crit} = \pm 2.306$) **21.** d **22.** c **23.** c.

Chapter 15

Introduction to the Analysis of Variance

CHAPTER OUTLINE

I. Introduction - *F* Distribution

A. Definition of *F*. *F* is the ratio of two independent population variance estimates.

B. Sampling distribution of *F*.

1. Take all possible values of size n_1 and n_2 from population.
2. Estimate population σ^2 from each of the samples using $s_1{}^2$ and $s_2{}^2$.
3. Calculate F_{obt} for all possible combinations of $s_1{}^2$ and $s_2{}^2$.
4. Calculate p(*F*) for each different value of F_{obt}.

C. *F* is always positive. Since *F* is a ratio of variance estimates, it will never be negative (i.e. squared numbers are always positive).

D. Shape of F distribution. F distribution is positively skewed.

E. Median value of F. With equal n's, median F value equals one.

F. Family of curves. F distribution is a family of curves for each combination of df.

G. Degrees of Freedom. Degrees of freedom for numerator $= n_1 - 1$; for denominator, df $= n_2 - 1$

II. ANOVA

A. Use.

1. Used to analyze data from experiments that employ more than two groups or conditions (can also be used for analyzing two conditions).
2. Used instead of many pairwise t tests in order to hold the probability of making a Type I error at α.

III. Overview of One-Way ANOVA

A. One overall comparison. F test allows us to make one overall comparison that tells whether there is a significant difference between the means of the groups.

B. Use. ANOVA can be used for

1. Independent groups design also called simple randomized group design or the one-way analysis of variance -- independent groups design or single factor experiment, independent groups design. In the independent groups design, there are as many groups as there are levels of the independent variable.

2. Repeated measures design.

C. Number of groups. In the independent groups design, there are as many groups as there are levels of the independent variable.

D. Hypothesis testing.

1. H_1 is nondirectional.
2. H_0 states that the different conditions are all equally effective, i.e., $\mu_1 = \mu_2 = \mu_3 = \cdots = \mu_k$

E. IV affects only the mean. ANOVA assumes that the independent variable affects only the mean of the scores but not the variance, i.e.,

$$_1^2 = {}_2^2 = {}_3^2 = \cdots \quad {}_k^2$$

F. Partioning of total variability. ANOVA partitions total variability of data (SS_T) into the variability that exists within each group (SS_W) and the variability between groups (SS_B). The SS stands for sum of squares.

G. Independent estimates of population variance. SS_B and SS_W are both used as independent estimates of the H_0 population variance.

H. F ratio:

$$F_{\text{obt}} = \frac{\textbf{between-groups variance estimate } (s_B^2)}{\textbf{within-groups variance estimate } (s_W^2)}$$

I. Underlying concept. s_B^2 increases with magnitude of effect of independent variable while s_W^2 is unaffected. The larger the F ratio the more unreasonable H_0 becomes.

J. Decision rule. If $F_{obt} \geq F_{crit}$, reject H_0.

IV. Within-Groups Variance Estimate, s_W^2

A. Analogous to s_W^2 for t test. Analogous to s_W^2 for t test but is for ≥ 2 groups. It is an estimate of σ^2

B. Conceptual equation:

$$s_W{}^2 = \frac{SS_1 + SS_2 + SS_3 + \cdots + SS_k}{N - k}$$

where $N = n_1 + n_2 + n_3 + \cdots + n_k$ and $N - k = \text{df}_W$

C. SS_W. SS_W is the within-groups sum of squares and is the numerator of s_W^2.

$$s_W{}^2 = SS_W/\text{df}_W$$

V. Between-Groups Variance Estimate, s_B^2

A. s_B^2 is an estimate of σ^2. If H_0 is true, the variance between the means of the samples is an estimate of σ^2.

B. s_B^2 is second estimate of σ^2. Using $s_{\overline{X}}{}^2$ as estimate for $\sigma_{\overline{X}}{}^2$, we can derive a second estimate of σ^2.

C. Conceptual equation.

$$s_B^2 = \frac{n\Sigma\left(\overline{X} - \overline{X}_G\right)^2}{k-1}$$

$$= \frac{n\left[(\overline{X}_1 - \overline{X}_G)^2 + (\overline{X}_2 - \overline{X}_G)^2 + (\overline{X}_3 - \overline{X}_G)^2 + \cdots + (\overline{X}_k - \overline{X}_G)^2\right]}{k-1}$$

where $\overline{X}_G$ is the overall mean of all the scores.

D. <u>Numerator and denominator of s_B^2</u>. The numerator of s_B^2 is called the between-groups sum of squares SS_B. The denominator is the df for s_B^2:

$$s_B^2 = SS_B/\text{df}_B$$

E. <u>Relationship between IV and s_B^2</u>. As the size of effect of the independent variable increases, $\Sigma\left(X - \overline{X}_G\right)^2$ increases, thereby increasing SS_B and s_B^2.

VI. *F* Ratio

A. <u>Relationship between s_B^2 and IV</u>. s_B^2 increases with size of effect of independent variable.

B. <u>Relationship between s_W^2 and IV</u>. Since independent variable affects only the mean not the variance, s_W^2 does not change.

C. <u>Relationship between *F* and IV</u>. Since $F = s_B^2/s_W^2$, as the independent variable effect increases, the *F* ratio increases.

D. <u>*F* values and real effect of IV</u>. If H_0 is true, *F* is expected to equal 1. The larger *F* becomes, the more reasonable it is that the independent variable has an effect.

E <u>Underlying concept</u>. **$F = (\sigma^2$ + effects of independent variable)$/\sigma^2$.**

F. <u>F < 1</u>. If $F < 1$, retain H_0.

G. <u>Relationship between *F* and *t*</u>. **$F = t^2$.** ANOVA can be used when $k = 2$ instead of *t* test.

VII. Calculating F_{obt}

A. <u>Calculate SS_B</u>:

$$SS_B = \left[\frac{\sum X_1^2}{n_1} + \frac{\sum X_2^2}{n_2} + \frac{\sum X_3^2}{n_3} + \cdots + \frac{\sum X_k^2}{n_k}\right] - \frac{\left(\sum^{\text{all scores}} X\right)^2}{N}$$

B. <u>Calculate SS_W</u>:

$$SS_W = \sum^{\text{all scores}} X^2 - \left[\frac{\sum X_1^2}{n_1} + \frac{\sum X_2^2}{n_2} + \frac{\sum X_3^2}{n_3} + \cdots + \frac{\sum X_k^2}{n_k}\right]$$

C. <u>Calculate SS_T as a check for work (remember $SS_T = SS_B + SS_W$)</u>.

$$SS_T = \sum^{\text{all scores}} X^2 - \frac{\left(\sum^{\text{all scores}} X\right)^2}{N}$$

D. Calculate df.

$$\mathbf{df}_B = k - 1$$
$$\mathbf{df}_W = N - k$$
$$\mathbf{df}_T = N - 1$$

E. Calculate s_B^2.

$$s_B^2 = SS_B/\mathrm{df}_B$$

F. Calculate s_W^2.

$$s_W^2 = SS_W/\mathrm{df}_W$$

G. Calculate F_{obt}.

$F_{obt} = s_B^2/s_W^2$, then evaluate with F_{crit}

VIII. Assumptions Underlying Use of ANOVA

A. Normally distributed. Populations from which samples are drawn are normally distributed.

B. Homogeneity of variance. Samples have equal variances (homogeneity of variance).

C. Violations of assumptions:

1. F is robust if sample sizes are equal.

2. F is minimally affected by violations of population normality.

IX. Size of Effect

A. Estimated Omega Squared ($\hat{\omega}^2$). With ANOVA, we use estimated omega squared ($\hat{\omega}^2$) to estimate the size of effect. Conceptually, very similar to r^2, the coefficient of determination. For the one-way, independent groups ANOVA,

$$\hat{\omega}^2 = \frac{SS_B - (k-1)\,s_W{}^2}{SS_T + s_W{}^2}$$

$\hat{\omega}^2$ gives a relatively unbiased estimate of effect size.

B. Eta Squared (η^2). A second method of estimating size of effect. Gives an upwards biased estimate. For the one-way independent groups ANOVA,

$$\eta^2 = \frac{SS_B}{SS_T}$$

Since η^2 gives a more biased estimate than $\hat{\omega}^2$, it is better to use $\hat{\omega}^2$.

X. Power of the Analysis of Variance

A. Sample size. Increasing sample size increases power.

B. Real effect of IV. The larger the real effect of the independent variable, the higher is the power.

C. Sample variability. The higher the sample variability, the lower is the power.

XI. Multiple Comparisons

A. Uses of multiple comparison techniques. When ANOVA is used on $k > 2$ groups, multiple comparisons are made to determine which conditions differ. The F ratio from the ANOVA tells whether some conditions do differ but not which ones.

B. *A priori* or planned comparisons.

1. These comparisons are planned in advance of the experiment and often arise from predictions based on theory or prior research.
2. Do not correct for higher probability of a Type I error.
3. Use t test for independent groups except we use s_W^2 as an estimate of σ^2 for the denominator of the t equation. Equation is:

$$t_{\text{obt}} = \frac{\overline{X}_1 - \overline{X}_2}{\sqrt{s_W^2\left(\frac{1}{n_1} + \frac{1}{n_2}\right)}}$$

with df $= N - k$.

If $n_1 = n_2 = n$, the formula simplifies to:

$$t_{\text{obt}} = \frac{\overline{X}_1 - \overline{X}_2}{\sqrt{\frac{2s_W^2}{n}}}$$

If $|t_{obt}| \geq |t_{crit}|$, reject H_0

4. This same *t* test can be applied for testing any pair of sample means that was planned prior to the experiment.
5. The number of planned comparisons should be kept to a minimum and flow logically from the experimental design.

C. A posteriori or post hoc comparisons.

1. General.

   a. Unplanned comparisons. These are comparisons between groups not planned prior to the experiment.
   b. Maintain Type I error rate at α. They maintain the Type I error rate at α while making all possible comparisons between sample means.
   c. Use *Q* or studentized range distributions. Sampling distributions for multiple comparisons are called *Q* or studentized range distributions and are developed by randomly taking *k* samples of equal *n* from the same population and determining the difference between the highest and lowest sample means.
   d. Error rates.

      1. Experiment-wise error rate is the probability of making one or more Type I errors for the full set of possible comparisons in an experiment.
      2. Comparison-wise error rate is the probability of making a Type I error for any of the possible comparisons.

2 Tukey HSD Test

a. Maintains the experiment-wise Type I error rate at α.

b. Statistic calculated is:

$$Q_{obt} = \frac{\overline{X}_i - \overline{X}_j}{\sqrt{\frac{s_W^2}{n}}}$$

where X_i = the larger of the two means being compared

X_j = the smaller of the two means being compared

$s_W 2$ = the within-groups variance estimate

n = the number of subjects in each group

c. Since $X_i > X_j$, Q_{obt} is always positive.

d. Use Q distribution to evaluate Q_{obt}.

e. Q_{crit} depends on n, k, and α.

If $Q_{obt} \geq Q_{crit}$, reject H_0.

f. Overall F must be significant to apply the HSD or any *post hoc* test.

2. Newman-Keuls Test

a. *Post hoc* test that holds the comparison-wise error rate at α rather than for the entire set of comparisons.

b. Q_{obt} is calculated the same way as for HSD test.

c. Varies the value of Q_{crit} for each comparison.

1. Value of Q_{crit} for any comparison is given by the sampling distribution of Q for the number of groups having means encompassed by $\overline{X}_i$ and $\overline{X}_j$ after all the means have been rank ordered.
2. The number of groups in the comparison is symbolized by r.
3. Different critical value for each comparison depending on r.

d. To use Newman-Keuls test the overall F must be significant.
e. Degrees of freedom equal $N - k$.
f. If $Q_{obt} \geq Q_{crit}$, reject H_0.
g. With unequal n's (provided they are not too different), use the harmonic mean (n) of the various n's and use it in the denominator of the Q equation for both HSD and Newman-Keuls.

$$\tilde{n} = \frac{k}{\frac{1}{n_1} + \frac{1}{n_2} + \frac{1}{n_3} + \cdots + \frac{1}{n_k}}$$

where $k =$ number of groups
$n_k =$ number of subjects in the kth group

D. Comparison of Tests.

1. Planned comparisons more powerful than *post hoc* tests since they do not correct for increased probability of making a Type I error.
2. Planned comparisons should be relatively few
3. Should flow logically from the experimental design
4. Newman-Keuls has higher experiment-wise Type I error rate than HSD.

5. Newman-Keuls is more powerful than HSD.
6. Newman-Keuls has a lower probability of making Type II error than HSD.
7. Which test to use depends on researcher's assessment of which error rate (Type I or II) to minimize.

CONCEPT REVIEW

Instead of using the mean for hypothesis

testing, the *F* test uses the (1) _______. (1) variance

The F statistic is the (2) _______ of two (2) ratio

independent variance estimates of the

same population variance. The *F* test

has a sampling distribution that gives

all possible values of (3) _______ (3) *F*

along with the (4) _______ for each (4) probability

value, assuming sampling is (5) _______ (5) random

from the population. The *F*

test has (6) _______ (6) two

values for degrees of freedom, one for each

variance estimate. The *F* ratio is always a

(7) _______ number. When *n*'s are	(7) positive
(8) _______ the (9) _______ value of *F*	(8) equal (9) median
will equal 1. Like the *t* test, there are a	
(10) _______ of curves for *F* depending on	(10) family
the different combinations of df_1 and df_2.	
The analysis of variance, abbreviated	
(11) _______ is appropriate to analyze data	(11) ANOVA
from experiments which use more than	
(12) _______ conditions or levels of the	(12) two
(13) _______ variable. In the ANOVA	(13) independ- ent
the independent variable is often referred to as	
a (14) _______. The ANOVA is used to	(14) factor
analyze results from an experiment instead	
of a series of *t* tests to avoid	
(15) _______ the probability of making a	(15) increas- ing
Type (16) _______ error. The use of the	(16) I
ANOVA (*F* test) allows us to make	

(17) _______ overall comparison between the	(17) one
(18) _______ of the groups. (Note: We are	(18) means
testing the (19) _______, not the	(19) means
(20) _______ of the groups.)	(20) variances
The simplest form of ANOVA is called the	
(21) _______ - (22) _______ ANOVA.	(21) one (22) way
It is used in an (23) _______ groups	(23) independ- ent
design. Though it is not required it is	
preferable to have equal(24) _______ in each	(24) *n*'s
group. The alternative hypothesis in the	
ANOVA is (25) _______.	(25) nondirec- tional
The one-way ANOVA (26) _______ the total	(26) partitions
variability, designated (27) _______ into two	(27) SS_T
sources; the variability that exists within	
each group, called the (28) _______-	(28) within
(29) _______ (30) _______ of (31) _______,	(29) groups (30) sum (31) squares
SS_W and the (32) _______ that exists	(32) variability

between the groups, called the (33) _______	(33) between
-groups (34) _______ of (35) _______, SS_B.	(34) sum (35) squares
Each sum of squares is used for an	
independent (36) _______ of the H_0 population	(36) estimate
(37) _______. The estimate based on the	(37) variance
within-groups variability is called the within	
-groups variance estimate, (38) _______.	(38) s_W^2
The estimate based on the between-groups	
variability is called the between-groups	
variance estimate, (39) _______. The *F* ratio	(39) s_B^2
, conceptually is:	
F_{obt} = (40) _______ / (41) _______	(40) s_B^2 (41) s_W^2
The (42) _______ -groups variance	(42) between
(43) _______ with the magnitude of the	(43) increases
effect of the independent variable while the	
(44) _______-groups variance is unaffected.	(44) within
Therefore, the (45) _______ the *F* ratio	(45) larger
the more (46) _______ the null	(46) unreason-

able

hypothesis becomes. To evaluate F_{obt} we

apply the same decision rules as usual, namely:

If (47) _______ ≥ (48) _______, reject H_0 (47) F_{obt} (48) F_{crit}

In this case we don't have to worry about a

(49) _______ value of F. If the value of (49) negative

F is (50) _______ 1, we do not have to (50) less than

bother to evaluate it since we know the

probability will be (51) _______ than α. (51) greater

The within-groups variance estimate is

determined the same way as for the t test

except we allow for more groups.

Conceptually:

$$s_W^{\ 2} = \frac{SS_1 + SS_2 + \cdots + (52)\,\_\_\_\_\_}{(n_1 - 1) + (n_2 - 1) + \cdots + (n_k - 1)}$$

(52) SS_k

where (53) _______ = the number of groups. (53) k

The numerator of s_W^2 above is called the

within-groups (54) _______ of (54) sum

(55) _______. For s_W^2 there are (55) squares

(56) _______ df. The computational formula for SS_W is:

(56) $N - k$

$$SS_W = (57)\_\_\_ - \left[\frac{(58)\_\_\_}{n_1} + \frac{(59)\_\_\_}{n_2} + \frac{(60)\_\_\_}{n_3} \right]$$

(57) $\sum^{\text{all scores}} X^2$

(58) $\left(\sum X_1\right)^2$

(59) $\left(\sum X_2\right)^2$

(60) $\left(\sum X_3\right)^2$

(61) $\left(\sum X_k\right)^2$

We can use SS_W in a simple formula for s_W^2:

s_W^2 = (62) _______ / (63) _______

(62) SS_W

(63) $N - k$ or df_W

The between-groups variance conceptual formula looks much like the formula for the sample variance:

$$s_B^{\,2} = \frac{(64)\_\_\_\_ \sum\left((65)\_\_\_\_ - (66)\_\_\_\_\right)^2}{(67)\_\_\_\_ - 1}$$

(64) n

(65) $\overline{X}$

(66) $\overline{X}_G$

(67) k

Expanding this equation gives the following conceptual formula:

$$s_B^{\,2} = \frac{n\left[\left(\overline{X}_1 - (68)\_\_\right)^2 + \left((69)\_\_ - (70)\_\_\right)^2 + \cdots + \left((71)\_\_ - (72)\_\_\right)^2\right]}{(73)\_\_ - 1}$$

(68) $\overline{X}_G$

(69) $\overline{X}_2$

(70) $\overline{X}_G$

(71) $\overline{X}_k$

(72) $\overline{X}_G$

(73) k

with $k - 1$ (74) _______ of (75) _______.

(74) degrees

(75) freedom

The numerator of the above equation is

(76) _______. $\overline{X}_G$ is the (77) _______ mean.

(76) SS_B

(77) grand or overall

One should now be able to see how s_B^2

(more specifically SS_B) (78) _______

(78) increases

as the effect of the independent variable

(79) _______.

(79) increases

One of the assumptions of the analysis

of variance is that the independent variable

affects only the (80) _______, not the

(80) mean

(81) _______, of each group. (82) _______

(81) variance

(82) s_W^2 or SS_W

does not change as a result of the independent variable. s_W^2 and s_B^2 are both independent estimates of (83) _______.

(83) σ^2

Therefore:

$$F_{obt} = s_B^2/s_W^2 = (\sigma^2 + (84)\ \_\_\_\_\_\_\_)/\sigma^2$$

(84) effects of the independent variable

The larger the value of F_{obt} the (85) _______ reasonable to assume H_0 is false.

(85) more

We stated earlier that the total variability or total sum of squares can be partitioned into (86) _______ parts in the one- way ANOVA. Therefore,

(86) two

$$SS_T = (87)\ \_\_\_\_\_\_\_ + (88)\ \_\_\_\_\_\_\_$$

(87) SS_W
(88) SS_B

Except to check our work (a very good idea) we only need to calculate two of the above and we can easily solve for the third quantity.

We can calculate SS_T as follows:

$$SS_T = \sum^{\text{all scores}} (89)\_\_\_\_\_ - \frac{(\sum^{\text{all scores}} (90)\_\_\_\_\_)^2}{(91)\_\_\_\_\_}$$

(89) X^2
(90) X
(91) N

For example, if $SS_T = 1920$ and $SS_B = 1805$,

SSW would equal (92) _______. The total — (92) 115

degrees of freedom equals (93) _______. — (93) $N - 1$

A slightly different way to look at the partitioning of the variability involves the deviation of individual score from the (94) _______ mean $(\overline{X} - \overline{X}_G)$. This deviation from the grand mean can be thought of as the deviation of the score from its own (95) _______ mean $(\overline{X} - \overline{X}_k)$, plus the deviation of that group from the (96) _______ mean $(\overline{X}_k - \overline{X}_G)$. The first deviation $(\overline{X} - \overline{X}_k)$ contributes to (97) _______ and the second $(\overline{X}_k - \overline{X}_G)$ contributes to (98) _______.

(94) grand
(95) group
(96) grand
(97) SS_W
(98) SS_B

As with the t test there are certain

assumptions to be considered. First, in	
using the ANOVA one assumes the populations	
from which the samples are drawn are	
(99) _______ distributed Second, one	(99) normally
assumes the samples are drawn from	
populations of (100) _______ variances.	(100) equal
Like the t test, the ANOVA is	
(101) _______. It is minimally affected by	(101) robust
violations of (102) _______ (103) _______.	(102) popula-tion (103) normality
It is also relatively insensitive	
to violations of (104) _______ of variance if	(104) homogen-eity
sample sizes in the groups are (105) _______	(105) equal
In the special case where k = (106) _______,	(106) 2
one can use either the ANOVA or the	
(107) _______ test. In this case, the	(107) t
following relationship will hold:	
F_{obt} = (108) _______	(108) t_{obt}^2

In addition to determining whether an effect is significant, it is often desirable to compute the (109) _______ of the effect. | (109) size

The (110) _______ of the total variability of Y | (110) proportion

that is accounted for by X is a measure of the (111) _______ of the effect of X on | (111) size

Y. As with the t test for independent groups, with one-way, Independent groups ANOVA , estimated omega squared, symbolized (112) _______ is used as a measure of the | (112) $\hat{\omega}^2$

(113) _______ of the effect. For one-way, | (113) size

Independent groups ANOVA,

$$\hat{\omega}^2 = \frac{(114)\_\_\_\_ - (k-1)(115)\_\_\_\_}{(116)\_\_\_\_ + (117)\_\_\_\_}$$

(114) SS_B
(115) $s_W^{\ 2}$
(116) SS_T
(117) $s_W^{\ 2}$

Another statistic used to measure the size of effect in one-way independent groups ANOVA is (118) _______. The equation | (118) η^2

for computing eta squared is,

$$\eta^2 = \frac{(119)\_\_\_\_\_}{(120)\_\_\_\_\_}$$

(119) SS_B
(120) SS_T

Unlike omega squared which gives an unbiased estimate of the effect of the Independent variable, eta squared gives an (121) _______ (122) _______ estimate.

(121) upward
(122) biased

In general with one-way, independent groups ANOVA, power can be affected by several factors. The (123) _______ the real effect of the independent variable, the higher the value of (124) _______.

(123) greater
(124) SS_B

The higher the value of SS_B, the higher the value of (125) _______ is. The higher the value of s_B^2, the (126) _______ the power is. Increasing the sample size will (127) _______ the value of s_B^2 and (128) _______ the value of s_W^2,

(125) s_B^2
(126) higher
(127) increase
(128) decrease

thereby (129) _______ power. As the	(129) increasing
sample variance increases, s_W^2 (130) _______,	(130) increases
causing power to (131) _______.	(131) decrease

Multiple Comparisons

So far we have learned that the ANOVA is a statistical technique to determine if the independent variable has had a significant effect in a multi-group experiment	
using (132) _____ overall test. But usually we are interested in determining which	(132) one
of the conditions (133) _______ from each	(133) differ
other. This is done by making (134) _______	(134) multiple
(135) _______ between pairs of	(135) compari-sons
group (136) _______. These comparisons may be either *a priori*, sometimes called	(136) means
(137) _______ comparisons or they	(137) planned
may be *a posteriori* or (138) _______	(138) post

(139) _______ comparisons.	(139) hoc
Planned comparisons are made in	
advance of the experiment. They are generally	
(140)_______ in (141)_______, and arise	(140) few (141) number
due to (142) _______. In the planned	(142) theory
comparisons we do not correct for the	
(143) _______ probability of a Type	(143) higher
(144) _______ error. There is controversy	(144) I
among statisticians about planned comparisons	
but for our purposes these comparisons do not	
need to be (145) _______ as long as there are	(145) independent or orthogonal
(146) _______ of them and they	(146) few
(147) _______ from the experimental	(147) flow
design logically.	
In doing planned comparisons we use	
the *t* test for (148) _______(149) _______	(148) independent (149) groups

except that we use (150) ______ (150) s_W^2

as an estimate of (151) ______ in the (151) σ^2

denominator. The equation becomes:

$$t_{\text{obt}} = \frac{\overline{X}_i - \overline{X}_j}{\sqrt{(152\ \_\_\_\_\_\left(\frac{1}{n_i} + \frac{1}{n_j}\right)}}$$

(152) $s_W^{\ 2}$

with (153) ______ degrees of freedom. (153) $N - k$

If $n_1 = n_2 = n$, then the equation simplifies

somewhat to:

$$t_{\text{obt}} = \frac{\overline{X}_i - \overline{X}_j}{\sqrt{\frac{(154)\ \_\_\_\_\_}{n}}}$$

(154) $2\,s_W^{\ 2}$

The significance of t_{obt} is evaluated in

the usual manner. If | (155) ______ | ≥ | (155) t_{obt}

(156) ______ |, then (157) ______ H_0. (156) t_{crit}
(157) reject

If comparisons are not planned a different

procedure referred to as (158)______ (158) post *hoc*

analysis is used. Since the comparisons

were not planned (159)_______ the experiment,	(159) before
we must correct for the (160) _______	(160) higher
probability of making a Type (161) _______	(161) I
error when comparisons are made for	
more than (162) _______ means. The	(162) two
sampling distributions for *post hoc*	
multiple comparisons are called the	
(163) _______ or (164) _______	(163) Q (164) student-zed
(165) _______ distributions. These	(165) range
distributions were developed by taking	
(166) ______samples of equal n from	(166) k
the (167) ______ population and	(167) same
determining the (168) ______ between the	(168) difference
highest and lowest sample (169) _______. The	(169) means
differences were then divided by (170) _______	(170) s_W2/n
producing distributions like the (171) _______	(171) t
distributions only these are appropriate for	

making (172) _______ comparisons.	(172) multiple
The two *a posteriori* tests described in this chapter are only two of many possible tests for multiple comparisons. The tests are different in how they correct for the increasing probability of making a Type I error.	
The (173) _______-wise error rate is the	(173) experi-ment
probability of making one or more Type I errors for the full set of possible comparisons in an experiment. The (174) _______-	(174) compari-son
wise error rate is the probability of making a Type I error for(175) _______ of the	(175) any
possible comparisons.	
The Tukey (176) _______ (177) _______	(176) honestly (177) significant
(178) _______ (HSD) test maintains the	(178) difference
(179) _______ -wise Type I error rate at	(179) experi-ment

(180) _______. The statistic calculated for	(180) α
this test is (181) _______. The formula is:	(181) Q
$$(182)\ \_\_\_\__{\text{obt}} = \frac{(183)\ \_\_\_\_ - (184)\ \_\_\_\_}{\sqrt{\dfrac{(185)\ \_\_\_\_}{(186)\ \_\_\_\_}}}$$	(182) Q (183) $\overline{X}_i$ (184) $\overline{X}_j$ (185) $s_W^{\ 2}$ (186) n
where	
$\overline{X}_i$ = the (187) _______ of the two means being compared.	(187) larger
$\overline{X}_j$ = the (188) _______ of the two means being compared.	(188) smaller
$s_W{}^2$ = the (189) _______ -groups (190) _______ estimate.	(189) within (190) variance
n = the (191) _______ of subjects in (192) _______ group	(191) number (192) each
To determine Q_{crit} we enter Table G using the degrees of freedom associated with (193) _______, (194) _______ (the number of groups in the experiment) and	(193) $s_W{}^2$ (194) k

(195) _______. Q_{obt} will always be	(195) α
(196) _______. If Q_{obt} (197) _______ Q_{crit}, reject H_0.	(196) positive (197) ≥
The (198) _______ -(199) _______ test is the other *post hoc* test presented. It is different from the HSD in that it maintains	(198) Newman (199) Keuls
the (200) _______-wise error rate at	(200) comparison
(201) _______. This means the Type I	(201) α
error rate is at α for (202) _______	(202) each
comparison, not for the (203) _______ set of comparisons. In order to accomplish this, the Newman-Keuls test varies the value of	(203) entire
(204) _______for each comparison. The value of Q_{crit} for any comparison is derived	(204) Q_{crit}
from the (205) _______ (206) _______	(205) sampling (206) distribution
of Q for the (207) _______ of groups which	(207) number

are encompassed by (208) _______	(208) $\overline{X}_i$
and (209) _______ after all the means	(209) $\overline{X}_j$
have been (210) _______ ordered. The	(210) rank
number of means encompassed by	
$\overline{X}_i$ and $\overline{X}_j$ is symbolized by (211) _______.	(211) *r*
The value of Q_{crit} depends on(212) _______,	(212) df
(213) _______, and (214) _______.	(213) *r* (214) α
Q_{obt} is calculated the same as for the	
HSD test substituting the appropriate values	
of $\overline{X}_i$ and $\overline{X}_j$. Just as for any *post hoc*	
comparison, to use the Newman-Keuls test	
the (215) _______ (216) _______ must be	(215) overall (216) *F*
(217) _______ before proceeding with	(217) significant
the multiple comparisons.	
Both of the above *a posteriori* tests are	
appropriate when there are (218) _______	(218) equal

n's in each group. If the *n*'s do not differ greatly we can still use these tests by calculating the (219) _______ (220) _______ of the various *n*'s and use it in the (221) _______ of the *Q* equation. The formula for the harmonic mean (symbolized by (222) _______) is:

(219) harmonic
(220) mean
(221) denominator
(222) $\tilde{n}$

$$\tilde{n} = \frac{(223)\ \_\_\_\_\_}{\frac{1}{n_1} + \frac{1}{n_2} + \frac{1}{n_3} + \cdots + \frac{1}{(224)\ \_\_\_\_\_}}$$

(223) k
(224) n_k

where

(225) _______ = number of groups

n_k = number of (226) _______ in the kth group

(225) k
(226) subjects

Since (227) _______ comparisons do not correct for an increased probability of making a Type I error, they are more (228) _______ than either of the *post hoc*

(227) planned
(228) powerful

tests. In - general, for the *a posteriori* tests, the (229) _______ test is	(229) Newman-Keuls
more powerful than the (230) _______ test.	(230) HSD
Newman-Keuls has a higher (231) _______-	(231) experiment
wise error rate, but a lower (232) _______	(232) Type II
error rate. One could say the HSD was the	
more (233) _______ of the two *a posteriori*	(233) conservative
tests. In deciding which test to use the	
researcher would choose the test which	
(234) _______ the appropriate error rate.	(234) minimized

EXERCISES

1. Why is SS_W insensitive to the effects of the independent variable?

2. Assuming that the sample sizes are equal, $N = 12$ and that $k = 3$, consider the following data:

$$s_1 = 8.113,\ s_2 = 8.816,\ s_3 = 7.258$$

a. What is the value of s_W2
b. What is the value of SS_W?

3. If $SS_T = 106$ and $SS_W = 24$, what is the value of SS_B?

4. A researcher is interested in knowing if different types of treatments have an effect on the number of relapses for clients with the diagnosis of alcoholism. He randomly selects a group of 24 clients from a waiting list at a local mental health center who consent to the study. Then he randomly assigns 6 subjects to each treatment group. Group 1 is a waiting list control group. They are given no treatment. Group 2 receives a medication that causes illness when alcohol is consumed. Group 3 gets cognitive behavior therapy, and Group 4 gets cognitive behavior therapy plus the drug. The researcher then records the number of relapses in the next 12 months. The following data are obtained.

Group 1	**Group2**	**Group 3**	**Group 4**
12	**4**	**7**	**1**
14	**6**	**5**	**3**
10	**6**	**5**	**0**
8	**6**	**2**	**2**
9	**3**	**1**	**2**
10	**2**	**1**	**1**

a. State the alternative hypothesis.
b. State the null hypothesis.
c. What is the value of SS_W?
d. What is the value of SS_B?
e. What is the value of SS_T?
f. What is the value of F_{obt}?
g. What do you conclude, using $\alpha = 0.05$?
h. What is the size of the effect, using $\hat{\omega}^2$?

i. What is the size of the effect, using ^{2}?

5. Is it possible to have a significant result with $k = 2$ using the t test, but when reanalyzing the data using the F test obtain a nonsignificant result? (Assume α is the same, the design is the same and you don't make a calculation error.)

6. The following data were obtained from an experiment to test the effect of different levels of variable A on heart rate (beats per minute).

	Levels of A	
Group 1	**Group 2**	**Group 3**
83	**79**	**90**
82	**86**	**80**
82	**72**	**70**
85	**73**	**65**
87	**70**	**79**

a. What is the value of s_B2?
b. What is the value of F_{crit}, $\alpha = 0.01$?
c. What is the value of F_{obt}?
d. What do you conclude using $\alpha = 0.01$?

7. A dentist wants to know if the flavor of toothpaste affects how often children brush their teeth, and therefore the number of cavities. She makes up 3 flavors of toothpaste, cherry, bubble gum and spinach and observes the following numbers of cavities at a 6-month check-up:

Cherry	Bubblegum	Spinach
7	5	8
2	2	9
8	6	6
5	4	2
3	1	9
3	8	5
0	4	3

a. State H_1.
b. State H_0.
c. What is the value of s_B2?
d. What is the value of s_W2?
e. What is the value of F_{crit}, $\alpha = 0.05$?
f. What is the value of F_{obt}?
g. What do you conclude using $\alpha = 0.05$?

8. Complete the following table where indicated?

Source	SS	df	s^2	F
Between	117.17	(b)__	(c)_____	(e)_____
Within	(a)_____	9	(d)_____	
Total	177.67	11		

9. Analyze the following data from an independent groups design using ANOVA.

Control	Experimental
50	45
60	38
51	40
52	48
51	50
49	40
50	44

a. What is the value of df_W?
b. What is the value of df_B?
c. What is the value of F_{obt}?
d. What would the value of t_{obt} be if one analyzed the data using the *t* test? Do a *t* test on the above data to confirm your results.
e. Would you reject H_0 using $\alpha = 0.05$?

10. Given the following data:

$\Sigma X_1 = 150$ $\Sigma X_2 = 98$ $\Sigma X_3 =$
85

$\Sigma X_1^2 = 4634$ $\Sigma X_2^2 = 1966$ $\Sigma X_3^2 =$
1575

$n_1 = 5$ $n_2 = 5$ $n_3 =$
5

a. What is the value of F_{obt}
b. What is the value of $\overline{X}_G$?
c. What is the value of SS_T?

11. Eight cities were asked to report data on fatalities in auto crashes. The Department of Safety wanted to know if safety

devices have an effect on mortality. The department received the following data on the number of fatalities per every 100 crashes where the vehicle was traveling over 40 m.p.h. The type of safety device was also reported. If the data were as follows, answer the questions below.

	Type of Device		
None	**Seat Belts**	**Airbags**	**Both**
16	15	14	6
12	11	9	6
8	7	5	4
10	10	10	5
6	5	4	3
2	1	0	3
9	10	8	6
9	8	6	5

a. What is the value of SS_B?

b. What is the value of $\overline{X}_G$?

c. What is the value of F_{obt}?

d. Would you reject H_0, using $\alpha = 0.01$?

12. Consider the following data from an independent groups design:

Group 1	Group 2	Group 3
6	5	20
7	4	92
9	4	68
8	3	31
2	1	4
3		10
		82
		212

a. Is it appropriate to use parametric ANOVA to analyze these results? Why or why not?

13. A pharmacologist is interested in determining whether or not three different psychoactive drugs differ in how long they remain in the body before they are excreted or metabolized. The pharmacologist randomly assigned 18 subjects to three different groups and recorded the number of days the drugs remained at measurable levels in each subject. The results are presented below.

Drug A	Drug B	Drug C
8	2	3
7	1	3
6	3	1
6	2	4
5	4	3
2	5	4

a. Is there a difference in the time it takes for these drugs to be cleared from the subjects? (Use $\alpha = 0.05$).
b. Which drugs are different in how long they remain in the subjects? (Use $\alpha = 0.05$ and the HSD test).
c. Same as 1b, but using the Newman-Keuls test.

14. A company that sells honey wishes to find out if the strain of bee they use (strain 1) is the most productive bee available. To study the question the four available strains are tested at their research farm to see how much honey is produced by each type of bee. Test hives are set up and the number of ounces of honey each hive produces is recorded and given below.

Strain 1	Strain 2	Strain 3	Strain 4
58	56	55	29
62	59	67	42
71	68	66	50
73	68	58	51
80	70	64	48

a. Are any *a priori* comparisons appropriate and if so which ones?
b. Using $\alpha = .05_{2\ \text{tail}}$ is there a difference between strain 1 and strain 2 using planned comparisons?
c. Using $\alpha = 0.05$ and the HSD test, which strains are different in the amount of honey they produce?
d. Same as 2c, using the Newman-Keuls test.

15. A clinical psychologist wants to know if there is any difference in the age of onset of paranoid schizophrenia, simple schizophrenia, or catatonic schizophrenia. The psychologist does a chart review and records the age of diagnosis for random samples of patients with the above diagnoses.

Simple	Paranoid	Catatonic
20	19	20
32	25	22
19	20	18
24	31	30
22	25	25

a. Is there any difference in the age of onset for these three types of schizophrenia? (use $\alpha = 0.01$).
b. Make all *post hoc* pairwise comparisons possible, using $\alpha = .01$ and the Newman-Keuls test.

16. A cancer researcher is interested in the effect of certain drugs on the growth rate of cancer cells. She grows 24 cultures of cancer cells and randomly divides them into four groups of 6 cultures each. Then she injects the cultures of each group with a different drug and counts the number of cells per square millimeter after 7 days. The following data are the results.

Control	Drug X	Drug Y	Drug Z
100	86	99	104
106	79	100	106
108	85	102	110
110	82	101	109
104	90	105	115
101	75	96	120

a. Use one-way ANOVA to determine if any of the drugs has an effect ($\alpha = 0.05$).
b. Use the Newman-Keuls test to compare the 4 drugs ($\alpha = 0.05$).

17. A physiological psychologist suspects that the sweat gland activity decreases in people as they get older. The psychologist believes that the highest readings should occur in the youngest people and the lowest readings should occur in the oldest people. The sweat gland activity is assessed by measuring the electrodermal activity of the skin. The following data are electrodermal readings measured in micromhos/cm .

Age Group		
Young	**Middle**	**Old**
8.6	8.5	2.1
11.2	9.0	1.0
10.1	10.2	3.6
9.0	7.6	5.4
9.8	9.5	2.7
13.6	7.0	2.6

a. Using an *a priori* test, evaluate whether or not the young group is different from the old group using $\alpha = 0.05_{1\ \text{tail}}$.
b. Using the Newman-Keuls test, evaluate all possible pair-wise comparisons using $\alpha = 0.05$. (Assume F_{obt} is significant.)
c. Would the results of part b have been different had we used the HSD test?

18. An experimental psychologist used 5 different methods to teach rats to run a maze. The results of an *F* test are shown below along with the sample means.

Source	SS	df	s^2	F
Between groups	208.8	4	52.2	7.25 *
Within groups	108.0	15	7.2	
Total	316.8	19		

*With $\alpha = 0.01$ $F_{crit} = 4.98$, therefore reject H_0.

$\overline{X}_1 = 13.75$, $\overline{X}_2 = 7.75$, $\overline{X}_3 = 13.25$, $\overline{X}_4 = 12.50$, $\overline{X}_5 = 5.75$ (seconds)

a. Do a planned comparison test on the difference between group 2 and group 4. Are the results significant using $\alpha = 0.05_{2\ tail}$? Assume equal *n*'s in each group.
b. What is the value of Q_{crit} at $\alpha = 0.01$ for the HSD test?
c. Is group 4 different from group 5 using $\alpha = 0.01$ with the Newman-Keuls test?

19. From an experiment the following data are obtained:

$\overline{X}_1 = 63.00$	$n_1 = 4$
$\overline{X}_2 = 73.25$	$n_2 = 4$
$\overline{X}_3 = 44.60$	$n_3 = 5$
$s_W^2 = 9.595$	df for $s_W^2 = 10$

a. Using $\alpha = 0.01$ and the Newman-Keuls test, compare the likelihood that $\overline{X}_1$ and $\overline{X}_3$ are random samples from the same population.
b. What is the value of *n*?

TRUE-FALSE QUESTIONS

T F 1. $s_B^2 + s_W^2 = s_T^2$.

T F 2. $SS_B + SS_W = SS_T$.

T F 3. A value of −1.0 for F_{obt} is never possible if the calculations are done correctly.

T F 4. The alternative hypothesis in an ANOVA can be either directional or nondirectional.

T F 5. The ANOVA is robust regardless of whether or not sample sizes are equal.

T F 6. A *t* test for independent groups can be used to analyze the results from an experiment with $k = 2$ groups.

T F 7. $F^2 = t$.

T F 8. A value of $F < 1.0$ will result in rejection of H_0.

T F 9. The *F* test for $k = 2$ is more powerful than the *t* test.

T F 10. It is always possible to partition the total variability in a simple ANOVA into two parts; the variability within groups and the variability between groups.

T F 11. An ANOVA can only be used when $n_1 = n_2 = \ldots = n_k$.

T F 12. The ANOVA is used to test the hypothesis that the group variances are equal.

T F 13. s_W^2 is sensitive to the effects of the independent variable.

T F 14. The sum of df_W and df_B equals $N - 1$.

T F 15. If F_{obt} is significant, it is possible to tell which groups differ from which without further analysis.

T F 16. If $F < 1.0$ we don't even have to bother to look up F_{crit} since the result cannot possibly be significant.

T F 17. If $n_1 \neq n_2 \neq n_k$, then $SS_T \neq SS_W + SS_B$.

T F 18. The analysis of variance is used to test for differences between sample means when $k \geq 2$.

T F 19. $F_{\mathrm{obt}} = s_W^2 / s_B^2$

T F 20. $F_{\mathrm{obt}} = (SS_B/\mathrm{df}_B)/(SS_W/\mathrm{df}_W)$

T F 21. $F_{\mathrm{obt}} = \dfrac{\sigma^2 + \text{the effects of the independent variable}}{\sigma^2 + \text{the effects of the dependent variable}}$

T F 22. The F distribution is a normal curve.

T F 23. With one-way, independent groups ANOVA, increasing sample variance decreases power.

T F 24. With one-way, independent groups ANOVA, increasing N causes s_B^2 to decrease, thereby producing an increase in power.

T F 25. The higher $\hat{\omega}^2$ is, the higher is power, other factors held constant.

T F 26. It is possible to have an overall F that is significant and have all the *a priori* comparisons be nonsignificant.

T F 27. It is not necessary to do an F test if one is only interested in doing planned comparisons between the groups.

T F 28. The most powerful and most legitimate way of doing multiple comparisons is to plan to do all possible pairwise comparisons prior to the experiment.

T F 29. If $k = 2$, the t test for independent groups, the F test, and the t test for planned comparisons are all essentially the same test.

T F 30. The reason we use s_W^2 based on all the groups instead of just two groups in doing multiple comparisons is because it gives us the best estimate of μ.

T F 31. If $k = 3$, Q_{crit} for the HSD is the same as Q_{crit} for Newman-Keuls if one is testing the difference between the highest and lowest means.

T F 32. The HSD test is less powerful than the planned comparisons.

T F 33. The Newman Keuls test has a lower probability of making a Type II error than the HSD test.

T F 34. The HSD holds the comparison-wise probability of making a Type I error at α.

T F 35. In the HSD test where $k = 3$ if the lowest and highest pair comparison is nonsignificant, none of the other comparisons will be significant either.

T F 36. All researchers agree that in order to do *a priori* comparisons the comparisons do not need to be orthogonal (independent).

T F 37. For the Newman-Keuls test as r increases so does the value of Q_{crit} if α and k are held constant.

T F 38. For both correlation and multiple comparisons r refers to the correlation coefficient.

T F 39. Q_{obt} can never be negative.

T F 40. If for $k = 4$, $n_1 = 10$, $n_2 = 12$, $n_3 = 30$, and $n_4 = 100$; one can still properly apply either the HSD or Newman-Keuls test provided one calculates the harmonic mean (n) to use in the denominator of the Q equation.

T F 41. In general, if the consequence of making a Type I error were much greater than making a Type II, it would be preferable to use the HSD test for a *post hoc* analysis.

T F 42. The probability of making a Type II error is the same, regardless of sample size, for the HSD test.

T F 43. If one did a series of *a priori* pairwise t tests on $k > 2$ groups and analyzed the results using the t distribution, there would be a substantially greater risk of making a Type II error than if one used HSD or Newman-Keuls tests and the Q distribution.

SELF-QUIZ

1. The total degrees of freedom for an experiment with $n_1 = 10$, $n_2 = 12$, and $n_3 = 10$ is ________.

   a. 2
   b. 32
   c. 31
   d. 29

2. Which of the following is(are) illegal values for F_{obt}?

   a. 1.00
   b. 0.96
   c. –2.97
   d. b and c

3. What is the value of F_{crit} for an experiment with $k = 3$, $n_1 = n_2 = n_3 = 10$ subjects, and $\alpha = 0.01$?

   a. 5.45
   b. 4.51
   c. 3.35
   d. 5.49

4. What is the value of F_{obt} for the following data?

Group 1	Group 2	Group 3
4	11	1
9	11	6
10	10	4

a. 5.67
b. 6.53
c. 1.95
d. 3.26

5. What is the value of F_{crit} for the data of question 4? Use $\alpha =$ 0.05.

a. 5.14
b. 4.76
c. 4.46
d. 4.07

6. What is your conclusion for the data of question 4? Again use $\alpha = 0.05$.

a. retain H_0
b. reject H_1
c. reject H_0
d. accept H_0

7. The estimated size of effect for the data of question 4 is ________.

a. 22.6% of the variability
b. 52.8% of the variability
c. 43.6% of the variability
d. 55.1% of the variability

8. What is the value of SS_T if $SS_B = 236$ and $SS_W = 54$?

a. 290
b. 182
c. 4.37
d. 100

9. By doing multiple t tests when there are more than 2 experimental groups we increase the risk of making what kind of mistake?

   a. accepting H_0
   b. Type I error
   c. Type II error
   d. all of the above

10. If $s_1^2 = 10$, $s_2^2 = 15$, and $s_3^2 = 12$, and $n_1 = n_2 = n_3$, then s_W^2 equals _______.

    a. cannot be determined from information given
    b. 37.0
    c. 12.33
    d. 3.51

11. If $s_1^2 = 9$, $s_2^2 = 6$, $s_3^2 = 8$, and $n_1 \neq n_2 \neq n_3$, then s_W^2 equals _______.

    a. cannot be determined from information given
    b. 7.667
    c. 2.77
    d. 23.0

12. If $\overline{X}_1 = 46$, $\overline{X}_2 = 50$, $\overline{X}_3 = 92$ and $n_1 = n_2 = n_3$, what is the value of $\overline{X}_G$?

    a. cannot be determined from information given
    b. 50.0
    c. 188.0
    d. 62.67

13. If $SS_W = 126$, $N = 28$, $k = 4$, then what is the value of s_W^2

a. cannot be determined from information given
b. 5.25
c. 42.0
d. 4.667

14. If $df_B = 3$ and $df_T = 29$ and $F_{obt} = 3.15$, what would you conclude using $\alpha = 0.05$?

a. reject H_0
b. reject H_1
c. retain H_0
d. retain H_1

15. If $k = 2$ and $t_{obt} = 2.95$, what would the value of F_{obt} be for an independent groups design?

a. 2.95
b. 8.70
c. 1.72
d. 0.05

16. If $s_B^2 = 27.9$ and $s_W^2 = 54.2$, what is the value of F_{obt}?

a. impossible result, there must be an error
b. 1.94
c. 0.51
d. 2.76

17. If $SS_T = 96$, $SS_W = 47$, and $SS_B = 68$, what would you conclude?

a. reject H_0
b. retain H_0

c. cannot be determined from information given
d. there must be an error in the calculations

18. Which of the following produce increases in power

a. increases in SS_B
b. increases in s_B^2
c. increases in SS_W
d. increases in s_W^2
e. increases in $\hat{\omega}^2$
f. a and b
g. a, b, and f

19. If the independent variable *X* accounts for a high percentage of the variability of *Y*, then we would expect $\hat{\omega}^2$ to be _______.

a. low
b. moderate
c. independent
d. high

20. The value for Q_{crit} using the HSD test for 6 groups with 6 subjects in each group and $\alpha = 0.01$ is _______.

a. 4.30
b. 4.23
c. 5.24
d. 5.37

21. The harmonic mean when $k = 3$ and $n_1 = 5$, $n_2 = 4$, and $n_3 = 7$ is _______.

a. 5.06
b. 1.69

c. 0.19
d. 0.06

22. In general the number of planned comparisons should ________.

    a. equal k
    b. be kept to a minimum
    c. equal n_k
    d. equal zero

23. Which of the following tests hold the experiment-wise error rate at α?

    a. HSD test
    b. Newman-Keuls
    c. F test
    d. a and c

24. In general, which of the following tests is the most powerful test to detect a difference between group means?

    a. a t test for a planned comparison
    b. the HSD test
    c. the Newman-Keuls test
    d. all equally powerful

25. If $n = 8$ and $s_W^2 = 18.2$, what is the value of the denominator for a planned comparison?

    a. 4.55
    b. 2.27
    c. 1.51
    d. 2.13

26. Using the Newman-Keuls test, if $\overline{X}_1 = 10.2$, $\overline{X}_2 = 16.1$, $\overline{X}_3 = 12.6$, and $\overline{X}_4 = 9.0$ the value of Q_{crit} for the comparison of $\overline{X}_1$ and $\overline{X}_4$ with $\alpha = 0.01$ is _______. s_W^2 has 12 degrees of freedom.

    a. 5.05
    b. 5.50
    c. 4.32
    d. 5.84

For problems 27 - 30 consider the following information from an experiment to compare differences between the length of time a house paint lasts before it begins to fade. The number of subjects are the same in each group.

Source	SS	df	s2	F
Between groups	**720.096**	**2**	**360.048**	**21.54**
Within groups	**300.857**	**18**	**16.714**	
Total	**1020.950**	**20**		

$\overline{X}_1 = 21.5714$, $\overline{X}_2 = 26.5714$, $\overline{X}_3 = 35.7143$, $\alpha = .01$

27. Using the HSD test the value of Q_{obt} for the testing the difference between $\overline{X}_1$ and $\overline{X}_2$ is _______.

    a. 5.92
    b. 2.29
    c. 3.24
    d. 1.05

28. The value of Q_{crit} for the *post hoc* comparison of $\overline{X}_1$ and $\overline{X}_2$ using the HSD test is _______ at $\alpha = 0.05$.

    a. 2.97
    b. 3.61
    c. 4.00
    d. 4.70

29. The value of Q_{crit} for the *post hoc* comparison of $\overline{X}_1$ and $\overline{X}_2$ using the Newman-Keuls test is _______ at $\alpha = 0.05$

    a. 2.97
    b. 4.07
    c. 3.61
    d. 4.00

30. The value of t_{obt} for the planned comparison of $\overline{X}_2$ and $\overline{X}_3$ is _______.

    a. 4.96
    b. 5.82
    c. 2.97
    d. 4.18

ANSWERS

Exercises: 1. Because the subjects within each group receive the same level of the independent variable, variability among the scores within each group cannot be due to differences in the effect of the independent variable.

2a. 65.41 **2b.** 588.66

3. 82

4a. The treatments have an effect on the number of relapses experienced by clients with alcoholism. **4b.** The treatments have no effect on the number of relapses experienced by clients with alcoholism. **4c.** 76 **4d.** 270 **4e.** 346 **4f.** F_{obt} = 23.68 **4g.** F_{crit} = 3.10; therefore reject H_0 **4h.** $\hat{\omega}^2 = 0.739$ **4i.** $^2 = 0.780$.

5. No, the tests are equivalent and equally powerful. With a little algebra one can derive the t equation from the F equation with k = 2

6a. 92.067 **6b.** 6.93 **6c.** 1.96 **6d.** F_{crit} = 6.93, F_{obt} = 1.96; retain H_0

7a. H_1 = the flavor of toothpaste has an effect on the number of cavities. **7b.** H_0 = the flavor of toothpaste has no effect on the number of cavities. **7c.** 8.190 **7d.** 7.190 **7e.** 3.55 **7f.** 1.14 **7g.** $F_{obt} < F_{crit}$; therefore retain H_0.

8a. 60.50 **8b.** 2 **8c.** 58.585 **8d.** 6.722 **8e.** 8.715

9a. 12 **9b.** 1 **9c.** 14.23 **9d.** 3.77 **9e.** F_{crit} = 4.75, reject H_0

10a. F_{obt} = 9.18 **10b.** 22.2 **10c.** 782.4

11a. 85.094 **11b.** 7.281 **11c.** 2.09 **11d.** F_{crit} = 4.57, therefore retain H_0

12. No, the homogeneity of variance assumption is violated. The F test is not robust with unequal n's.

13a. $F_{obt} = 5.24$; $F_{obt} = 3.68$; therefore, reject H_0. There is a difference in the time required for these drugs to be cleared from the body. **13b.** Q_{obt} for Drug *A* and Drug *B* comparison = 4.09; $Q_{crit} = 3.67$; therefore reject H_0. Q_{obt} for Drug *A* and Drug *C* comparison =3.83; $Q_{crit} = 3.67$; therefore reject H_0. Q_{obt} for Drug *B* and Drug *C* comparison = 0.26; $Q_{crit} = 3.67$; therefore retain H_0. **13c.** All values for Q_{obt} will be the same as for the HSD test only Q_{crit} will change. For comparison of Drug *A* and *B*, $Q_{crit} = 3.67$; therefore reject H_0. For comparison of Drug *A* and *C*, $Q_{crit} = 3.01$; therefore reject H_0. For comparison of Drug *B* and *C*, $Q_{crit} = 3.01$; therefore retain H_0. By both analyses Drug *A* differs from both drugs *B* and *C* in the time required to be cleared from the body. However, we cannot detect a difference between drugs *B* and *C*.

14a. Yes, *a priori* comparisons of strain 1 and 2, 1 and 3, 1 and 4 are appropriate. **14b.** $t_{obt} = 0.97$; $t_{crit} = \pm 2.120$ therefore retain H_0. **14c.** Q_{crit} for all comparisons = 4.05. Q_{obt} for 1 vs 2 = 1.37; retain H_0. Q_{obt} for 1 vs 3 = 2.02; retain H_0. Q_{obt} for 1 vs 4 = 7.36; reject H_0. Q_{obt} for 2 vs 3 = 0.65; retain H_0. Q_{obt} for 2 vs 4 = 6.00; reject H_0. Q_{obt} for 3 vs 4 = 5.34; reject H_0. **14d.** Q_{obt} will be same as for 2d only Q_{crit} will change. Q_{crit} for 1 vs 2 = 3.00; retain H_0. Q_{crit} for 1 vs 3 = 3.65; retain H_0. Q_{crit} for 1 vs 4 = 4.05; reject H_0. Q_{crit} for 2 vs 3 = 3.00; retain H_0. Q_{crit} for 2 vs 4 = 3.65; reject H_0. Q_{crit} for 3 vs 4 = 3.00; reject H_0

15a. $F_{obt} = 0.053$; retain H_0. We cannot detect a difference here. **15b.** No *post hoc* comparisons are significant.

16a. $F_{obt} = 39.21$; F_{crit} for 3 and 20 df = 3.10; therefore reject H_0. **16b.** Q_{obt} (control vs *X*) = 11.48; $Q_{crit} = 3.58$; reject H_0. Q_{obt} (control vs *Y*) = 2.26; $Q_{crit} = 2.95$; retain H_0. Q_{obt} (control vs *Z*) = 3.04; $Q_{crit} = 2.95$; reject H_0. Q_{obt} (*X* vs *Y*) = 9.22; $Q_{crit} = 2.95$; reject H_0. Q_{obt} (*X* vs *Z*) = 14.53; $Q_{crit} = 3.96$; reject H_0. Q_{obt} for *Y* vs Z = 5.31; Q_{crit} = 3.58; retain H_0. Drug X differs significantly from all the others.

17a. t_{obt} = 8.52; t_{crit} = 1.753; therefore reject H_0. **17b.** Q_{obt} (young vs middle) = 2.82; Q_{crit} = 3.01; retain H_0. Q_{obt} (young vs old) = 12.04; Q_{crit} = 3.67; reject H_0. Q_{obt} (middle vs old) = 9.23; Q_{crit} = 3.01; reject H_0. **17c.** No, since Q_{crit} would be 3.67 for all tests if we had used the HSD test.

18a. t_{obt} = 2.50; t_{crit} = 2.131; therefore reject H_0. **18b.** Q_{crit} = 5.56. **18c.** Q_{obt} = 5.031; Q_{crit} = 4.84; therefore reject H_0.
19a. Q_{obt} = 12.30; Q_{crit} = 4.48; therefore reject H_0. **19b.** 4.29.

True-False: **1.** F **2.** T **3.** T **4.** F **5.** F **6.** T **7.** F **8.** F **9.** F **10.** T **11.** F **12.** F **13.** F **14.** T **15.** F **16.** T **17.** F **18.** T **19.** F **20.** T **21.** F **22.** F **23.** T **24.** F **25.** T. **26.** F **27.** T **28.** F **29.** T **30.** F **31.** T **32.** T **33.** T **34.** F **35.** T **36.** F **37.** T **38.** F **39.** T **40.** F **41.** T **42.** F **43** F.

Self-Quiz: **1.** c **2.** c **3.** d **4.** b **5.** a **6.** c **7.** d **8.** a **9.** b **10.** c **11.** a **12.** d **13.** b **14.** a (F_{crit} = 2.98) **15.** b **16.** c **17.** d **18.** g **19.** d. **20.** c **21.** a **22.** b **23.** d **24.** a **25.** d **26.** c **27.** c **28.** b **29.** a **30.** d.

Chapter 16

Introduction to Two-Way Analysis of Variance

CHAPTER OUTLINE

I. Qualitative Analysis of Two-Way ANOVA

A. General points.

1. Two-Way ANOVA is a more sophisticated technique used to investigate more than one factor (or independent variable).

2. A factorial experiment is one in which the effect of two or more factors are assessed in one experiment. In a factorial experiment the treatments used are combinations of the levels of both factors.

3. Allows us to evaluate the effect of two independent variables (e.g. *A* and *B*) and the interaction between them (A x *B*).

4. Main effects. The effect of Factor *A* averaged over the levels of Factor *B* and the effect of Factor *B* averaged over the levels of Factor *A* are called main effects.

5. Interaction effects. An interaction effect exists when the effect of one factor is not the same at all levels of the other factor.

B. Analyzing data.

1. Calculate four variance estimates:

a. s_W^2; the within-groups variance estimate
b. s_R^2; row variance estimate (sensitive to *A* effects)
c. s_C^2; column variance estimate (sensitive to *B* effects)
d. s_{RC}^2 ; row by column variance estimate (sensitive to interaction effects)

2. <u>Test for effects</u>.

For *A*: $\mathbf{F_{obt} = s_R^2/s_W^2}$

For *B*: $\mathbf{F_{obt} = s_C^2/s_W^2}$

For interaction between *A* and *B*: $\mathbf{F_{obt} = s_{RC}^2/s_W^2}$

II. Quantitative Analysis of Two-Way ANOVA

A. <u>Steps in performing the two-way ANOVA</u>.

1. The total sum of squares (SS_T) is partitioned into four components: the within-cells sum of squares (SS_W), the row sum of squares (SS_R), the column sum of squares (SS_C), and the row × column sum of squares (SS_{RC}).

2. Four variance estimates of σ^2 are formed by dividing each of the above four sum of squares by their degrees of freedom. These estimates are the within-cells variance estimate ($s_W{}^2$), the row variance estimate ($s_R{}^2$), the column variance estimate ($s_C{}^2$), and the row × column variance estimate ($s_{RC}{}^2$).
3. Three *F* ratios are formed from the four variance estimates. The *F* ratios are:

$\mathbf{F_{obt} = s_R{}^2/s_W{}^2}$ *Main effect of variable A*

$\mathbf{F_{obt} = s_C{}^2/s_W{}^2}$ *Main effect of variable B*
$\mathbf{F_{obt} = s_{RC}{}^2/s_W{}^2}$ *Interaction effect of variables A and B*

4. The *F* ratios are evaluated using the following decision rule.

$$\textbf{If } F_{obt} \geq F_{crit}\textbf{, reject } H_0$$

III. Within-cells Variance Estimate ($s_W{}^2$)

A. Equivalent to $s_W{}^2$ in one-way ANOVA. Provides an estimate of σ^2. $s_W{}^2$ is a measure or the inherent variability of the scores from subject-to-subject. It is based on the variability of the scores within each cell. Therefore, it does not reflect any treatment effect.

B. Equations.

1. Within-cells variance estimate ($s_W{}^2$).

$$s_W{}^2 = SS_W/\text{df}_W$$

where SS_W = within-cells sum of squares
df_W = within-cells degrees of freedom

2. Within-cells sum of squares (SS_W).

$$SS_W = SS_{11} + SS_{12} + \quad + SS_{rc}$$ *Conceptual equation*

where SS_{11} = sum of squares for the cell of row 1 and column 1

SS_{rc} = sum of squares for the cell of row r and column c

$$SS_W = \sum^{\text{all scores}} X^2 - \left[\frac{\left(\sum^{\text{cell 11}} X\right)^2 + \left(\sum^{\text{cell 12}} X\right)^2 + \cdots + \left(\sum^{\text{cell } rc} X\right)^2}{n_{\text{cell}}} \right]$$ *Computational equation*

where $\left(\sum^{\text{cell } rc} X\right)^2$ = sum of the scores in the cell of row r and column c, squared

3. Within-cells degrees of freedom (df_W).

$$df_W = rc(n - 1)$$

where r = number of rows
c = number of columns

IV. Row Variance Estimate (s_R^2)

A. Used to assess the main effect of variable *A*. Based on the differences between row means. Analogous to s_B^2 in one-way ANOVA. It is an estimate of σ^2 + the effects of variable *A*, averaged over the levels of variable *B*. If variable *A* has no effect, then the population row means are equal. Hence:

$$\mu_{a_1} = \mu_{a_2} = \cdots = \mu_{a_r}$$

and the differences among sample row means is just due to random sampling from identical populations.

B. Equations.

1. Row variance estimate (s_R^2).

$$s_R^2 = SS_R/df_R$$

where SS_R = row sum of squares
df_R = row degrees of freedom

2. Row sum of squares (SS_R).

$$SS_R = n_{\text{row}}\left[\left(\overline{X}_{\text{row }1} - \overline{X}_G\right)^2 + \left(\overline{X}_{\text{row }2} - \overline{X}_G\right)^2 + \cdots + \left(\overline{X}_{\text{row }r} - \overline{X}_G\right)^2\right]$$ *Conceptual equation*

where $$\overline{X}_{\text{row 1}} = \frac{\sum\limits^{\text{row 1}} X}{n_{\text{row 1}}}$$

$$\overline{X}_{\text{row r}} = \frac{\sum\limits^{\text{row r}} X}{n_{\text{row r}}}$$

$$\overline{X}_G = \text{grand mean} = \frac{\sum\limits^{\text{all scores}} X}{N}$$

$$SS_R = \left[\frac{(\overset{\text{row 1}}{\Sigma} X)^2 + (\overset{\text{row 2}}{\Sigma} X)^2 + \cdots + (\overset{\text{row } r}{\Sigma} X)^2}{n_{\text{row}}} \right] - \frac{(\overset{\text{all scores}}{\Sigma} X)^2}{N}$$ *Computational equation*

3. Row degrees of freedom (df_R).

$$\mathbf{df_R = r - 1}$$

V. Column Variance Estimate (s_C^2).

A. Used to assess the main effect of variable *B*. Based on the differences between column means. Analogous to s_B^2 in one-way ANOVA. It is an estimate of σ^2 + the effects of variable *B*, averaged over the levels of variable *A*. If variable *B* has no effect, then the population column means are equal. Hence:

$$\mu_{b_1} = \mu_{b_2} = \cdots = \mu_{b_c}$$

and the differences among the sample column means are due to random sampling form identical populations.

B. Equations:

1. Column variance estimate (s_C^2).

$$\mathbf{s_C^2 = SS_C/df_C}$$

where SS_C = column sum of squares
df_C = column degrees of freedom

2. Column sum of squares (SS_C).

$$SS_C = n_{\text{col.}}\left[\left(\overline{X}_{\text{col.}1} - \overline{X}_G\right)^2 + \left(\overline{X}_{\text{col.}2} - \overline{X}_G\right)^2 + \cdots + \left(\overline{X}_{\text{col.}c} - \overline{X}_G\right)^2\right]$$

Conceptual equation

where $$\overline{X}_{\text{col. }1} = \frac{\sum\limits^{\text{col. }1} X}{n_{\text{col. }1}}$$

$$\overline{X}_{\text{col. }c} = \frac{\sum\limits^{\text{col. }c} X}{n_{\text{col. }c}}$$

$$SS_C = \left[\frac{\left(\sum\limits^{\text{col. }1} X\right)^2 + \left(\sum\limits^{\text{col. }2} X\right)^2 + \cdots + \left(\sum\limits^{\text{col. }c} X\right)^2}{n_{\text{col.}}}\right] - \frac{\left(\sum\limits^{\text{all scores}} X\right)^2}{N}$$

Computational equation

3. Column degrees of freedom (df_C).

$$df_C = c - 1$$

VI. Row × Column Variance Estimate (s_{RC}^2).

A. Used to assess the interaction effect of variables *A* and *B*. An interaction exists when the effect of one of the variables is not the same at all levels of the other variable. Based on the differences among the cell means beyond that which is predicted by the individual effects of the two variables. If there is no interaction and any main effects are removed, then the population cell means are equal. Hence:

$$\mu_{a_1b_1} = \mu_{a_1b_2} = \cdots = \mu_{a_rb_c}$$

and differences among cell means must be due to random sampling from identical populations.

B. Equations.

1. Row × column variance estimate (s_{RC}^2).

$$s_{RC}^2 = SS_{RC}/\mathrm{df}_{RC}$$

where SS_{RC} = row × column sum of squares
df_{RC} = row × column degrees of freedom

2. Row × column sum of squares (SS_{RC}).

$$SS_{RC} = n_{\text{cell}}\left[\left(\overline{X}_{\text{cell 11}} - \overline{X}_G\right)^2 + \left(\overline{X}_{\text{cell 12}} - \overline{X}_G\right)^2 + \cdots + \left(\overline{X}_{\text{cell rc}} - \overline{X}_G\right)^2\right] - SS_R - SS_C$$

Conceptual equation

where
$$\overline{X}_{\text{cell 11}} = \frac{\sum^{\text{cell 11}} X}{n_{\text{cell}}}$$

$$\overline{X}_{\text{cell } rc} = \frac{\sum^{\text{cell } rc} X}{n_{\text{cell}}}$$

$$SS_{RC} = \left[\frac{\left(\sum^{\text{cell 11}} X\right)^2 + \left(\sum^{\text{cell 12}} X\right)^2 + \cdots + \left(\sum^{\text{cell } rc} X\right)^2}{n_{\text{cell}}}\right] - \frac{\left(\sum^{\text{all scores}} X\right)^2}{N} - SS_R - SS_C$$

Computational equation

3. Row × column degrees of freedom (df_{RC}).

$$df_{RC} = (r - 1)(c - 1)$$

VII. Computing *F* Ratios.

A. Equations.

1. To test the main effect of variable *A* (row effect):

$$F_{obt} = s_R^2 / s_W^2$$

2. To test the main effect of variable *B* (column effect):

$$F_{obt} = s_C^2 / s_W^2$$

3. To test the interaction effect of variables *A* and *B* (row × column effect):

$$F_{obt} = s_{RC}^2 / s_W^2$$

B. Decision rule.

$$\text{If } F_{obt} \geq F_{crit}, \text{ reject } H_0$$

This decision rule applies for all three F_{obt} values. Note that it is possible to have all three comparisons significant; none of the comparisons significant; and all possible combinations between these two extremes.

VIII. Interpreting the F comparisons

A. Three *F* values to interpret. As with one-way ANOVA, a significant *F* value indicates a real effect. *A* significant *F* involving $s_R{}^2$ indicates the *A* variable has had a real main effect. A significant *F* involving $s_C{}^2$ indicates the *B* variable has had a real main effect. A significant *F* involving $s_{RC}{}^2$ indicates there is a real interaction between *A* and *B*. The effects of one or both variables are not the same at all levels of the other variable. If *F* is not significant, we cannot conclude there has been a real effect, and we must retain H_0. Whichever way we conclude, it is possible we have made a Type I or II error.

IX. Multiple Comparisons

A. Similar to multiple comparisons discussed in Chapter 15. When conducting a two-way ANOVA, the experimenter is usually interested in more than just determining the main and interaction effects. There is an interest in determining which levels of which variables are having real effects. This is conceptually similar to Chapter 15 where we took up multiple comparisons in conjunction with one-way ANOVA. However, in the two-way ANOVA it is more complicated and beyond the scope of this textbook.

X. Assumptions Underlying Two-Way ANOVA

A. Populations normally distributed. The populations from which the samples were taken are normally distributed.

B. Homogeneity of variance. Similar to homogeneity of variance discussed in Chapter 15.

C. Violations of Assumptions. Two-way ANOVA is robust with regard to violations of these assumptions, provided the samples are of equal size

CONCEPT REVIEW
Qualitative Analysis of Two-Way ANOVA

A slightly more complicated way of analyzing experiments is the two-way ANOVA. The one-way ANOVA examines one	
independent variable or (1) _______	(1) factor
with several (2) _______ of that independent variable.	(2) levels
The two-way ANOVA allows us to evaluate	
in (3) _______ experiment the	(3) one
effect of (4) _______ independent variables	(4) two
and the (5) _______ between them.	(5) interaction
Such experiments are called	
(6) _______ experiments. In a factorial experiment the treatments used are	(6) factorial
(7) _______ of the levels of both	(7) combinations
factors. In such a design the effect of Factor	

A averaged over the

levels of Factor *B*, and the effect of Factor *B*

averaged over the

levels of Factor A are called(8) _______ (8) main

(9) _______. An (9) effects

(10) _______ effect occurs when the effect (10) interaction

of one factor is not the (11) _______ at all levels (11) same

of the other factor.

In analyzing data from a two-way ANOVA,

we determine

(12) _______ variance estimates. (12) four

They are (13) _______, (13) s_W^2

(14) _______, (15) _______, and (14) s_R^2
(15) s_C^2

(16) _______. The estimate s_W^2 is the (16) s_{RC}^2

within-(17) _______ variance estimate (17) cell

and is similar to s_W^2 in the simple one-way

ANOVA. It becomes the standard against

which the other estimates are

(18) _______. s_R^2 is called | (18) compared

the (19) _______ variance | (19) row

estimate and it is sensitive to the effects of variable

(20) _______. s_C^2 is the (21) _______ | (20) A
(21) column

variance estimate and it is sensitive to the effects

of variable (22) _______. | (22) *B*

These are both (23) _______ effects. | (23) main

The estimate (24) _______ is the (25) _______ | (24) s_{RC}^2
(25) row

by (26) _______ variance | (26) column

estimate. It is sensitive to the(27) _______ | (27) interaction

effects of variables *A* and *B*. The following

tests can be made.

For variable *A*:

F_{obt} = (28) _______$/s_W^2$ | (28) s_R^2

For variable *B*:

F_{obt} = (29) _______$/s_W^2$ | (29) s_C^2

For the interaction between *A* and *B*:

F_{obt} = (30) _______$/s_W^2$ | (30) s_{RC}^2

The values of F_{obt} are evaluated against

appropriate values of (31) _______. The main advantage of the two-way ANOVA is that we can do (32) _______ one-way experiments plus we are able to evaluate the (33) _______ between two (34) _______ variables.

(31) F_{crit}
(32) two
(33) interaction
(34) independent

Quantitative Analysis of Two-Way Anova

In analyzing data from a two-way ANOVA, the total sum of squares, symbolized by (35) _______ is partitioned into (36) _______ parts, the within sum of squares, symbolized by (37) _______, the row sum of squares, symbolized by (38) _______, the column sum of squares, symbolized by (39) _______, and the row × column sum of squares, symbolized by (40) _______.

(35) SS_T
(36) four
(37) SS_W
(38) SS_R
(39) SS_C
(40) SS_{RC}

Thus,

$SS_T = SS_W +$ (41) ____ + (42) ____ + (43) ____

(41) SS_R

(42) SS_C
(43) SS_{RC}

Each of the sum of squares,

SS_W, SS_R, SS_C, and SS_{RC} is used to form a

Variance used to form a variance estimate

of (44) _______, the variance (44) σ^2

of the null hypothesis population.

The variance estimates are derived by dividing

the sum of squares by its (45) _______. (45) degrees of

Thus,

$s_W^2 = SS_W/$(46) _______ (46) df_W

$s_R^2 = SS_R/$(47) _______ (47) df_R

$s_C^2 = SS_C/$(48) _______ (48) df_C

$s_{RC}^2 = SS_{RC}/$(49) _______ (49) df_{RC}

The within-cells variance estimate, symbolized by

(50) _______ is based on the differences (50) s_W^2

among scores (51) _______ each cell and (51) within

hence is not sensitive to the effects of

variables *A* and *B*. Thus, it is a measure

of σ^2 (52) _______ (52) uninfluenced

by the effects of *A* and *B*. It is analogous to (53) _______ in the one-way ANOVA. It is the standard against which we assess the effects of (54) _______.

(53) s_W^2

(54) *A* and *B*

The row variance estimate, symbolized by (55) _______ is based on the differences among the row means. Therefore, it is a measure of σ^2 (56) _______ the effects of variable *A*. If the F ratio involving s_R^2 is significant, we conclude that variable A has had a real average effect. We say that there is a (57) _______ effect for variable *A*.

(55) s_R^2

(56) plus

(57) main

The column variance estimate, symbolized by (58) _______ is based on the differences among the column means. Therefore it is a measure of σ^2 plus the effects of variable (59) _______. If the *F* ratio involving s_C^2 equals

(58) s_C^2

(59) *B*

or exceeds (60) _______ we reject H_0, (60) F_{crit}

and conclude that their was a real

(61) _______ effect for variable *B*. (61) main

The row × column variance estimate,

symbolized by (62) _______ is based on the (62) s_{RC}^2

differences among (63) _______ (63) cell

means beyond that predicted by the individual

effects of the two

variables. It is used to evaluate the (64) _______ (64) interaction

between variables *A* and *B*. A

significant *F* ratio here indicates that the

effect of one or both of the variables

is (65) _______ at all levels (65) not the same

of the other variable.

Computing the F ratios

The within-cells sum of squares

(SS_W) is just the (66) _______ within each (66) *SS*

cell added together. Thus,

$SS_W = SS_{11} + SS_{12} + \cdots +$ (67) _______ (67) SS_{rc}

The computational equation for SS_W is:

$$SS_W = \sum^{\text{all scores}} X^2 - \left[\frac{\left(\sum^{\text{cell } 11} X\right)^2 + \left(\sum^{\text{cell } 12} X\right)^2 + \cdots + \left(\sum^{\text{cell } rc} X\right)^2}{n_{\text{cell}}} \right]$$

where $\left(\sum^{\text{cell } rc} X\right)^2$ = sum of the scores in

the cell or row r and

column c, (68) _______ (68) squared

The within-cells degrees of freedom is

given by:

$df_W =$ (69) _____ (69) $rc(n-1)$

The conceptual equation for the row

sum of squares is:

$$SS_R = n_{\text{row}}\left[\left(\overline{X}_{\text{row }1} - \overline{X}_G\right)^2 + \left(\overline{X}_{\text{row }2} - \overline{X}_G\right)^2 + \cdots + \left(\overline{X}_{\text{row }r} - \overline{X}_G\right)^2\right]$$

As the effect of the A variable increases, $\left(\overline{X}_{\text{row }r} - \overline{X}_G\right)^2$ also

(70) _______. Thus SS_R is sensitive (70) increases

to the effect of variable

(71) _______. The computational (71) *A*

equation for SS_R is:

$$SS_R = \left[\frac{(\overset{\text{row}}{\underset{1}{\Sigma}} X)^2 + (\overset{\text{row}}{\underset{2}{\Sigma}} X)^2 + \cdots + (\overset{\text{row}}{\underset{r}{\Sigma}} X)^2}{n_{\text{row}}} \right] - \frac{(\overset{\text{all}}{\underset{\text{scores}}{\Sigma}} X)^2}{N}$$

where $(\overset{\text{row}}{\underset{r}{\Sigma}} X)^2$ indicates summing all the scores in row *r*

and then (72) _______. (72) squaring

The row degrees of freedom is given by:

$df_R =$ (73) _______ (73) $r - 1$

SS_C is very much like SS_R. However,

it involves (74) _______ means rather (74) column

than row means. Hence it is sensitive to

the effects of variable *B*. This can be best seen

from the conceptual equation for SS_C.

$$SS_C = n_{\text{col.}} \left[(\overline{X}_{\text{col.1}} - \overline{X}_G)^2 + (\overline{X}_{\text{col.2}} - \overline{X}_G)^2 + \cdots + (\overline{X}_{\text{col.c}} - \overline{X}_G)^2 \right]$$

As the effect of the B variable increases, $(\overline{X}_{\text{col.c}} - \overline{X}_G)$ also

(75) _______. Thus SS_C is sensitive | (75) increases

to the effect of variable (76) _______. | (76) *B*

The computational equation for SS_C is:

$$SS_C = \left[\frac{\left(\sum^{\text{col. }1} X\right)^2 + \left(\sum^{\text{col. }2} X\right)^2 + \cdots + \left(\sum^{\text{col. }c} X\right)^2}{n_{\text{col.}}} \right] - \frac{\left(\sum^{\text{all scores}} X\right)^2}{N}$$

where $\left(\sum^{\text{col. }c} X\right)^2$ indicates summing all

the scores in col. *c* and then (77) _______. | (77) squaring

The column degrees of freedom is given by:

df_C = (78) _______ | (78) $c - 1$

The row × column sum of squares

(SS_{RC}) is a measure of the differences

among (79) _______ means when the individual | (79) cell

effects of variables A and B have

been (80) _______. Thus, it measures | (80) removed

the (81) _______ between the two variables. | (81) interaction

The computational equation for SS_{RC} is:

$$SS_{RC} = \left[\frac{(\overset{\text{cell 11}}{\Sigma} X)^2 + (\overset{\text{cell 12}}{\Sigma} X)^2 + \cdots + (\overset{\text{cell } rc}{\Sigma} X)^2}{n_{\text{cell}}} \right] - \frac{(\overset{\text{all scores}}{\Sigma} X)^2}{N}$$

$$-SS_R - SS_C$$

where $(\overset{\text{cell } rc}{\Sigma} X)^2$ indicates summing all the scores in cell *rc* and then (82) ______. (82) squaring

The row × column degrees of freedom is given by:

df_{RC} = (83) ______ (83) $(r-1)(c-1)$

To compute the various *F* ratios, we first compute the four (84)______ and check our computations by computing (85) ______. The equation for computing SS_T scores is:

(84) sum of squares

(85) SS_T

$$SS_T = (86) \_\_\_\_\_\_ - (87) \_\_\_\_\_\_$$

(86) $\overset{\text{all scores}}{\Sigma} X^2$

(87) $\frac{(\overset{\text{all scores}}{\Sigma} X)^2}{N}$

The equation for checking our *SS*

computations is:

SS_T = (88) ____ + (89) ____ + (90) ____ + (91) ____ (88) SS_R
(89) SS_C
(90) SS_{RC}
(91) SS_W

Once we are sure our SS computations are correct, we compute the four degrees of freedom. Then we compute the four

(92) _______. Next we compute the (92) variance estimates

three (93) _______ (93) *F* ratios

Finally, we evaluate each *F* ratio by using the decision rule,

If F_{obt} (94) _______ F_{crit}, reject H_0 (94) ≥

A significant *F* value indicates a

(95) _______ effect. A (95) real

significant *F* involving s_R^2 indicates

the (96) _______ variable has (96) *A*

had a real main effect. A significant

F involving s_C^2 indicates the (97) _______ (97) *B*

variable had a real (98) _______ effect. A (98) main

significant *F* involving $s_{RC}2$ indicates there

is a real (99) _______ between *A* and *B* (99) interaction

If *F* is not significant, we must

(100) _______ H_0. Whichever way (100) retain

we conclude, it is possible we have made

a (101) ______ or (102) ______ error. (101) Type I
(102) Type II

To validly use the *F* test in analyzing the

data in the two-way ANOVA, the populations

from which the samples were taken

should be (103) _______ distributed,

and there should be (103) normally

(104) _______ of variance. (104) homogeneity

The *F* test is (105) _______ with (105) robust

regard to violations of these mathematical assumptions.

EXERCISES

1. An experiment was conducted to assess the effects of a minor tranquilizer on a performance task at different levels of stress. The levels of stress (Factor *A*) were moderate and

high and the levels of tranquilizer (Factor *B*) were none and moderate. A two-way ANOVA was done on the data and there was a significant *A* x *B* interaction. Explain what this would mean.

2. An independent groups experiment is conducted involving variables *A* and *B*. There are three levels of each variable, and four subjects randomly assigned to each cell. The following data is obtained:

	B		
A	**Level 1**	**Level 2**	**Level 2**
Level 1	2 5 4 7	3 5 6 7	10 8 6 7
Level 2	3 2 4 6	2 6 5 4	8 8 9 6
Level 3	4 7 9 7	5 8 10 7	6 9 7 10

a. What are the null hypotheses for this experiment?
b. Using the two-way ANOVA with $\alpha = 0.05$, what do you conclude?

3. A sleep researcher is interested in determining whether drinking coffee affects ability to sleep. The researcher has a hunch that it does, but that the effect will depend on how close to bedtime the coffee is drunk. An experiment is

conducted in which there are 3 levels of the amount of coffee which is drunk (variable *B*) and two levels of when the coffee is drunk (variable *A*). The levels of amount drunk are one, two and three cups of coffee and the coffee is drunk either in the morning or two hours before bedtime. The dependent variable is the number of hours slept that night. Thirty subjects are run in the experiment, with five each being randomly assigned to each cell. The following data are collected.

	Coffee		
Time	**1 Cup**	**2 Cups**	**3 Cups**
	8	9	7
	7	8	9
Morning	8	9	8
	6	6	9
	9	7	8
	6	5	5
	7	6	4
2 hrs before bedtime	8	6	6
	8	6	4
	9	7	5

What do you conclude? Analyze with two-way ANOVA and α = 0.05.

4. A researcher working with elderly people is interested in whether memory decreases with age. An independent groups experiment is conducted with three age groups, 30 - 39, 40 - 49, and 50 - 59. Two memory tasks are given, a difficult, long passage to be memorized and a relatively easy, shorter passage. Subjects are randomly assigned, four to

each cell. Scores are percent correct, with 100 being a perfect score. The data are shown here:

Memory Task	Age (Years)		
	30 - 39	40 - 49	50 - 59
Easy	90	88	85
	95	92	94
	97	96	89
	93	91	91
Difficult	80	77	67
	84	75	66
	82	79	69
	79	74	71

a. What are the null hypotheses for this experiment?
b. Analyze these results using two-way ANOVA and $\alpha = 0.05$. What do you conclude?

5. Complete the following table. There are 5 subjects per cell.

Source	SS	df	s^2	F_{obt}
Rows	1082.358	2		
Columns	2158.364	3		
Row $\times$ Column	785.228			
Within-cells	3290.875			
Total				

6. A scientist, interested in weight regulation, believes that if exercise in increased, appetite will increase and more food will be eaten. If this is true, it could reduce the benefits of exercise on weight reduction. A 2 $\times$ 3 factorial experiment is conducted to investigate the effect of exercise on food intake. The experiment involves three levels of exercise and uses

male and female rats as subjects. Daily food consumption (grams) is monitored as the dependent variable. The animals are allowed to eat as much food as desired. The following data (grams/day) are obtained:

	Exercise		
Gender	**Low**	**Moderate**	**High**
	8	12	14
	10	13	15
Female	11	11	18
	9	15	15
	10	12	14
	13	19	21
	15	12	15
Male	14	15	16
	14	18	20
	13	17	22

Analyze these results with two-way ANOVA, using $\alpha = 0.01$. What do you conclude?

TRUE-FALSE QUESTIONS

T F 1. $s_T^2 = s_R^2 + s_C^2 + s_{RC}^2 + s_W^2$.

T F 2. $SS_T = SS_R + SS_C + SS_{RC} + SS_W$.

T F 3. In general, the alternative hypothesis for two-way ANOVA can be directional.

T F 4 s_W^2 is sensitive to the effects of variable *A* or *B*.

T F 5. s_C^2 is a measure of only σ^2, regardless of whether the variables have a main effect.

T F 6. s_R^2 is sensitive to the effects of variable A.

T F 7. The sum of df_R, df_C, df_{RC}, and df_W is $N - 1$.

T F 8. Two-way ANOVA is like two one-way ANOVA experiments, plus it gives us the ability to analyze the interaction of the two variables.

T F 9. A 2×4 factorial experiment has 2 levels of one variable and 4 levels of the other variable.

T F 10. It is not possible to have an interaction effect unless one of the variables also has a main effect.

T F 11. A main effect refers to the average effect of the variable.

T F 12. In a two-way ANOVA, we do three F tests.

T F 13. s_W^2 is the standard for evaluating the main and interaction effects.

T F 14. In two-way ANOVA, the variability within each cell is used to produce an estimate of σ^2 that is independent of any treatment effects.

T F 15. The variability of cell means is used as a basis of both the interaction effect and the within-cell variance estimate.

T F 16. The assumptions underlying two-way ANOVA and one-way ANOVA are different.

T F 17. In general, when an experimenter does a two-way ANOVA, he has no interest in which levels of the variable have a real effect.

T F 18. In two-way ANOVA, it is possible for all or none, or any combination in between, of the *F* values to be significant.

T F 19. An assumption of two-way ANOVA is that the sample cell variances are equal.

T F 20. An assumption of two-way ANOVA is that the sample column and row scores are normally distributed.

SELF-QUIZ

For questions 1 - 9, use the following data, collected from an independent groups design. $\alpha = 0.05$.

	Variable B		
Variable A	**1**	**2**	**3**
	5	4	4
	2	6	9
1	1	3	7
	4	1	5
	2	3	4
	6	6	5
	2	7	7
2	3	4	5
	3	3	4
	2	3	8

1. The value of F_{obt} for evaluating the row effect is _______.

   a. 0.66
   b. 6.23
   c. 0.29
   d. 5.28

2. The value of F_{crit} for evaluating the row effect is _______.

   a. 3.40
   b. 7.82
   c. 4.26
   d. 5.61

3. The conclusion regarding the main effect of variable *A* is _______.

   a. Retain H_0. We cannot conclude variable *A* has main effect.
   b. Accept H_0. We cannot conclude variable *A* has main effect.
   c. Reject H_0. Variable *A* has a significant main effect.
   d. Reject H_0. Variable *A* has no effect.

4. The value of F_{obt} for evaluating the column effect is _______.

   a. 0.66
   b. 6.23
   c. 0.29
   d. 5.28

5. The value of F_{crit} for evaluating the column effect is _______.

   a. 3.40
   b. 7.82
   c. 4.26
   d. 5.61

6. The conclusion regarding the main effect of variable *B* is _______.

   a. Retain H_0. We cannot conclude variable *B* has main effect.
   b. Accept H_0. We cannot conclude variable *B* has main effect.
   c. Reject H_0. Variable *B* has a significant main effect.
   d. Reject H_0. Variable *B* has no effect.

7. The value of F_{obt} for evaluating the row × column effect is _______.

   a. 0.66
   b. 6.23
   c. 0.29
   d. 5.28
8. The value of F_{crit} for evaluating the row × column effect is _______.

   a. 3.40
   b. 7.82
   c. 4.26
   d. 5.61
9. The conclusion regarding the interaction effect of variables *A* and *B* is _______.

   a. Retain H_0. We cannot conclude there is a significant interaction
   b. Accept H_0. There is no interaction effect between *A* and *B*
   c. Reject H_0. There is a significant interaction effect.
   d. Reject H_0. There is no interaction effect between *A* and *B*

10. Consider the following graphic results from a 2 × 2 factorial experiment. These results show _______.

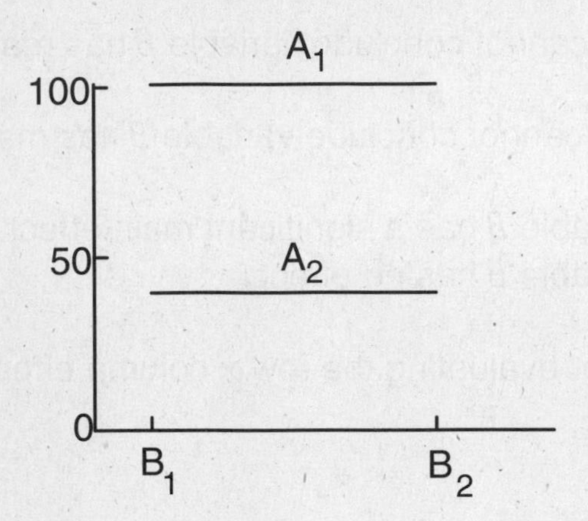

a. there are no significant main effects or interaction effects
b. there is a significant main effect for factor *A*, no other significant effects
c. there is a significant main effect for factor *B*, no other significant effects
d. there is a significant interaction effect, no other significant effects

11. Consider the following graphic results from a 2 × 2 factorial experiment. These results show _______.

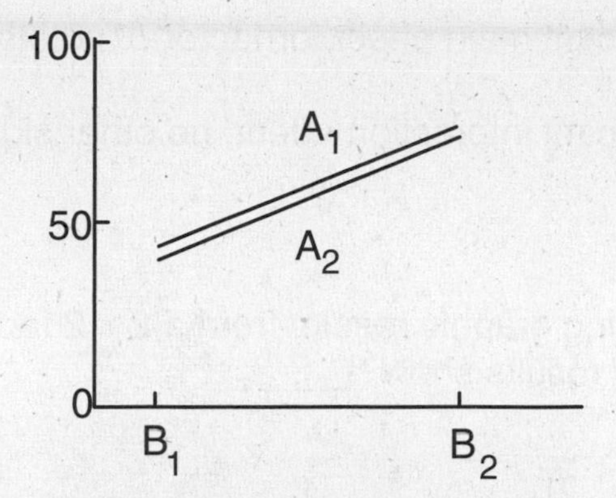

a. there are no significant main effects or interaction effects
b. there is a significant main effect for factor *A*, no other significant effects
c. there is a significant main effect for factor *B*, no other significant effects
d. there is a significant interaction effect, no other significant effects

12. Consider the following graphic results from a 2×2 factorial experiment. These results show _______.

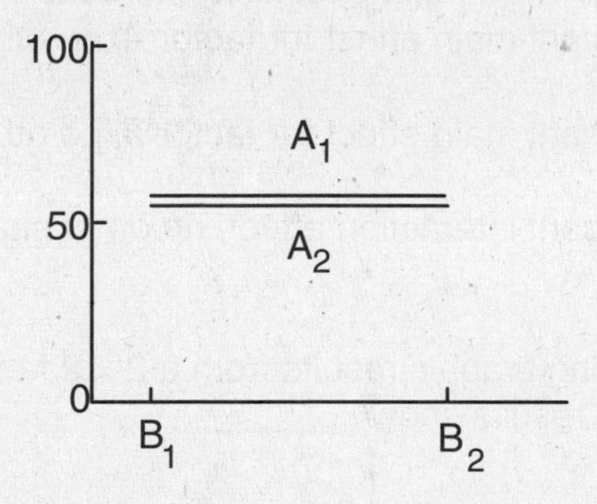

a. there are no significant main effects or interaction effects
b. there is a significant main effect for factor *A*, no other significant effects

c. there is a significant main effect for factor *B*, no other significant effects
d. there is a significant interaction effect, no other significant effects

13. Consider the following graphic results from a 2 × 2 factorial experiment. These results show _______.

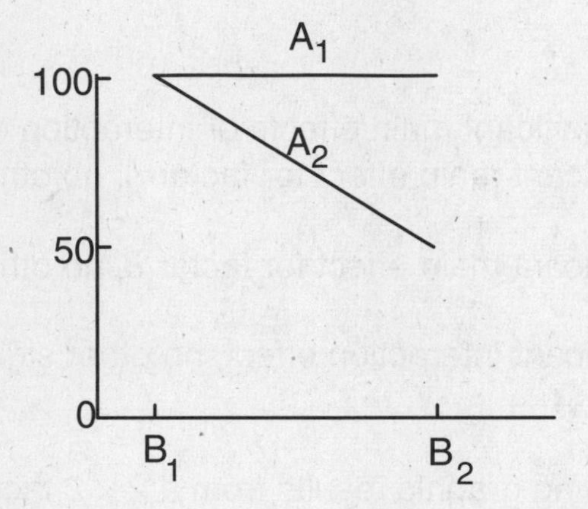

a. there are no significant main effects or interaction effects
b. there is a significant main effect for factor *A*, no other significant effects
c. there is a significant main effect for factor *B*, no other significant effects
d. there is a significant interaction effect, no other significant effects

14. Consider the following graphic results from a 2 × 2 factorial experiment. These results show _______.

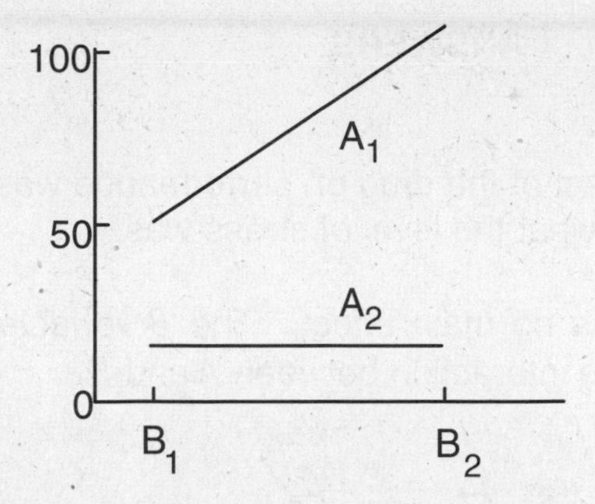

a. there is a significant main effect for factor *A*, no other significant effects
b. there is a significant main effect for factor *B*, no other significant effects
c. there is a significant interaction effect, no other significant effects
d there is a significant main effect for factor *A*, a significant interaction effect, and no other significant effects
e. there is a significant main effect for factor *B*, a significant interaction effect, and no other significant effects

15. How many variance estimates are there in a 2 x 2 factorial design?

a. 1
b. 2
c. 3
d. 4

ANSWERS

Exercises: 1. The effect of the drug on performance was different depending on what the level of stress was.

2a. The *A* variable has no main effect. The *B* variable has no main effect. There is no interaction between *A* and *B*.

$$\mu_{a_1} = \mu_{a_2} = \cdots = \mu_{a_r}$$

$$\mu_{b_1} = \mu_{b_2} = \cdots = \mu_{b_c}$$

$$\mu_{a_1 b_1} = \mu_{a_1 b_2} = \cdots = \mu_{a_r b_c}$$

2b.

Source	SS	df	s^2	F_{obt}	F_{crit}
Rows (variable A)	30.167	2	15.083	4.60	3.35
Columns (Variable B)	52.667	2	26.334	8.03	3.35
Row × Column	11.667	4	2.917	0.89	2.73
Within-cells	88.50	27	3.278		
Total	183	35			

There is a significant main effect for variable *A* and variable *B*. The interaction effect is not significant. Therefore, we reject H_0 for both main effects, but retain H_0 for the interaction between *A* and *B*.

3.

Source	SS	df	s^2	F_{obt}	F_{crit}
Rows (Time)	22.533	1	22.533	21.81	4.26
Columns (Coffee amount)	6.200	2	3.100	3.00	3.40
Row × Column	14.467	2	7.233	7.00	3.40
Within-cells	24.800	24	1.033		
Total	68	29			

There is a significant main effect for rows and a significant row × column interaction. There is no significant columns effects. Thus drinking coffee in the morning had no apparent effect., no matter how much was drunk (up to 3 cups, of course). Drinking coffee 2 hours before bedtime interfered with sleep, and the more coffee drunk at this time, the greater the interference.

4a. Memory is the same for the easy and difficult tasks (main effect of variable A). There is no memory change with age (main effect of variable *B*). There is no interaction between age and memory of material of different levels of difficulty (*A* × *B* interaction).

$$\mu_{a_1} = \mu_{a_2} = \cdots = \mu_{a_r}$$

$$\mu_{b_1} = \mu_{b_2} = \cdots = \mu_{b_c}$$

$$\mu_{a_1 b_1} = \mu_{a_1 b_2} = \cdots = \mu_{a_r b_c}$$

4b.

Source	SS	df	s^2	F_{obt}	F_{crit}
Rows (Task difficulty)	1633.500	1	1633.500	200.70	4.41
Columns (Age)	292.000	2	146.000	17.94	3.55
Row × Column	84.000	2	42.000	5.16	3.55
Within-cells	146.500	18	8.139		
Total	2156	23			

There is a significant main effect for task difficulty. There is a significant main effect for age, and there is a significant interaction between age and task difficulty. The "difficult" material was harder to memorize independent of age differences; as age increased there was a loss in memory independent of task difficulty; and the age decrement was more pronounced with the difficult task.

5.

Source	SS	df	s^2	F_{obt}
Rows	1082.358	2	541.179	7.89
Columns	2158.364	3	719.455	10.49
Row × Column	785.228	6	130.871	1.91
Within-cells	3290.875	48	68.560	
Total	7316.825	59		

6.

Source	SS	df	s^2	F_{obt}	F_{crit}
Rows (Female/male)	108.300	1	108.300	26.63	7.82
Columns (Exercise)	140.467	2	70.233	17.27	5.61
Row × column	0.600	2	0.300	0.07	5.61
Within-cells	97.600	24	4.067		
Total	346.967	29			

There is a significant main effect for both rows and columns. The interaction effect is not significant. Thus, the amount of food females and males eat daily is significantly different, independent of exercise level. Increasing the level of exercise increases the amount of food eaten, independent of gender. Exercise level does not appear to have a different effect on males and females regarding the amount of food eaten.

True-False: **1.** F **2.** T **3.** F **4.** F **5.** F **6.** T **7.** T **8.** T **9.** T **10.** F **11.** T **12.** T **13.** T **14.** T **15.** F **16.** F **17.** F **18.** T **19.** F **20.** F.

Self-Quiz: **1.** a **2.** c **3.** a **4.** b **5.** a **6.** c **7.** c **8.** a **9.** a **10.** b **11.** c **12.** a **13.** d **14.** d. **15.** d.

Chapter 17

Chi-Square and Other Nonparametric Tests

CHAPTER OUTLINE

I. Distinctions Between Parametric and Nonparametric Tests

A. Parametric tests. Parametric tests (e.g., *t*, *z*, *F*) depend substantially on population characteristics or parameters for their use.

B. Nonparametric tests. Nonparametric tests (e.g., sign test) depend minimally on population characteristics.

C. Distribution free tests. Whereas parametric tests may require that samples be random from normally distributed populations, nonparametrics require that samples be random from populations with the same distributions, hence the term distribution free tests.

D. Advantages for parametric tests.

1. Parametric tests are generally more powerful and versatile.

2. Parametric tests are generally robust to violations of the test assumptions.

E. Examples of nonparametric tests.

1. Sign test
2. Mann-Whitney U test
3. Chi-square test
4. Wilcoxon matched-pairs signed ranks test
5. Kruskal-Wallis test

II. Chi-Square (χ^2) Single Variable Experiments

A. Use. Often used with nominal data.

B. What is tested. Tests if the observed results differ significantly from the results expected if H_0 were true.

C. Computational formula.

$$\chi^2_{obt} = \Sigma \frac{(f_o - f_e)^2}{f_e}$$

where f_o = the observed frequency in the cell
f_e = the expected frequency in the cell (if H_0 were true)
Σ = summation over all cells

D. Evaluation of χ^2_{obt}.

1. Family of curves
2. Vary with df
3. Lower df curves are positively skewed

4. $k-1$ degrees of freedom where k equals the number of groups or categories
5. The larger the discrepancy between the observed and expected results the larger the value of χ^2_{obt} and therefore the more unreasonable that H_0 is true.
6. If $\chi^2_{obt} \geq \chi^2_{crit}$, reject H_0

III. Chi-square: Test of Independence Between Two Variables

A. Use. Used to determine whether two variables are related.

B. Contingency table. This is a two-way table showing the contingency between two variables where the variables have been classified into mutually exclusive categories and the cell entries are frequencies.

C. Null Hypothesis. Null hypothesis states that the observed frequencies are due to random sampling from a population in which the proportions in each category of one variable are the same for each category of the other variable.

D. Alternative Hypothesis. Alternative hypothesis. Alternative hypothesis is that these proportions are different.

E. Calculation of χ^2 for contingency tables.

$$\chi^2_{obt} = \sum \frac{(f_o - f_e)^2}{f_e}$$

1. f_e can be found by multiplying the marginals (i.e. row and column totals lying outside the table) and dividing by N.
2. Sum $(f_o - f_e)^2/f_e$ for each cell.

F. Evaluation of χ^2_{obt}.

1. Degrees of freedom for experiments involving the contingency between two variables are equal to the number of f_o scores that are free to vary while at the same time keeping the column and row marginals the same. In equation form:

$$\mathbf{df} = (r - 1)(c - 1)$$

where r = number of rows in the contingency table
c = number of columns in the contingency table

2. If $\chi^2_{obt} \geq \chi^2_{crit}$, reject H_0

G. Assumptions underlying χ^2

1. Independence exists between each observation in the contingency table.

2. Sample size is large enough so that the expected frequency in each cell is at least 5 for tables where r or c is greater than 2.
3. If table is 1 x 2 or 2 x 2 then each expected frequency should be at least 10.

4. χ^2 can be used with any type of scaling if the data are reduced to mutually exclusive categories and frequency entries.

IV. Wilcoxon Matched-Pairs Signed Ranks Test

A. Use.

1. Used in correlated groups designs with data that are at least of ordinal scaling.
2. Used when assumptions of *t* test for correlated groups are seriously violated.

B. Power. Relatively powerful. More powerful than sign test, less powerful than *t* test.

C. Data. Considers both magnitude and direction of the rank order of the difference scores.

D. Alternative Hypothesis. Alternative hypothesis stated with no population parameters; e.g. independent variable affects dependent variable.

E. Null Hypothesis. Null hypothesis stated with no population parameters; e.g. independent variable has no effect on dependent variable.

F. Calculation of statistic T_{obt}.

1. Calculate the difference between each pair of scores.
2. Rank the absolute values of the difference scores from the smallest to the largest.
3. Assign to the resulting ranks the sign of the difference score whose absolute value yielded that rank.
4. Compute the sum of the ranks separately for the positive and negative signed ranks. The lower sum is T_{obt}.

5. As a check, the sum of the unsigned ranks should equal $n(n + 1)/2$.
6. If rows scores are tied such that the difference of the paired scores equals zero, then these scores are discarded and N reduced by one.

7. If ties occur in the difference scores, the ranks are given a value equal to the mean of the tied ranks.

G. Evaluation of T_{obt}.

1. **If $T_{obt} < T_{crit}$, reject H_0.** T_{crit} depends on α and N.

H. Assumptions of the signed ranks test.

1. Raw scores must be of at least ordinal scaling.
2. Difference scores must also be of at least ordinal scaling.

V. Mann-Whitney *U* Test

A. Use. Used in a two group, independent groups design as a substitute for the t test when its assumptions are seriously violated. Measures the degree of separation between the two sets of sample scores.

B. Requirements. It is a nonparametric test that requires only ordinal scaling of the dependent variable. Does not require population normality.

C. Analysis. Rank orders the scores, computes the sum of ranks for each group, and tests whether these sums are significantly different. Makes no prediction about population means.

D. Calculation of U_{obt} or U'_{obt}. Computes the statistic U_{obt} or U'_{obt}. To calculate U_{obt} or U'_{obt}

1. Combine all the scores and rank order them, beginning with 1 for the lowest score.
2. Sum the ranks for each group.
3. Substitute these values into the equations and compute U_{obt} and U'_{obt}. U_{obt} is always the smaller of the two results.
4. Equations:

$$U_{obt} = n_1 n_2 + \frac{n_1(n_1+1)}{2} - R_1$$

$$U_{obt} = n_1 n_2 + \frac{n_2(n_2+1)}{2} - R_2$$

where n_1 = Number of scores in group 1
n_2 = Number of scores in group 2
R_1 = sum of the ranks for group 1
R_2 = sum of the ranks for group 2

E. Evaluation of U_{obt}. Since U_{obt} *and* U'_{obt} give the same information regarding degree of separation, it is only necessary to evaluate one of them. The textbook always evaluates U_{obt}.

If $U_{obt} \le U_{crit}$, reject H_0

with U_{crit} found in Tables C.1-C.4 using α, n_1 and n_2. U_{crit} is the upper of the two entries found in the appropriate cell of the appropriate table.

VI. Kruskal-Wallis Test

A. Use. Used in independent groups design as a substitute for parametric ANOVA when its assumptions are seriously violated. Like parametric ANOVA, Kruskal-Wallis is a nondirectional test.

B. Requirements. It is a nonparametric test which requires only ordinal scaling of the dependent variable. Does not require population normality.

C. Analysis. Computes the sum of ranks for each group and tests whether these sums are significantly different. Makes no prediction about population means.

D. Calculation of statistic H_{obt}. Statistic computed is H_{obt}. To compute H_{obt}

1. Combine all the scores and rank order them, beginning with 1 for the lowest score.
2. Sum the ranks for each group.
3. Substitute these values into the equation and compute H_{obt}.
4. Equation:

$$H_{obt} = \left[\frac{12}{N(N-1)}\right]\left[\frac{R_1^2}{n_1} + \frac{R_2^2}{n_2} + \frac{R_3^2}{n_3} + \cdots + \frac{R_k^2}{n_k}\right] - 3(N+1)$$

where R_1 = sum of the ranks for sample 1
R_2 = sum of the ranks for sample 2
R_3 = sum of the ranks for sample 3
R_k = sum of the ranks for sample k
k = number of samples or groups

E. Evaluation of H_{obt}.

If $H_{obt} \geq H_{crit}$, reject H_0

with H_{crit} found in Table H using df = $k - 1$.

CONCEPT REVIEW

A (1) _______ inference test is one that depends substantially on population characteristics for its use. Tests that require only minimal knowledge about the population characteristics are called (2) _______ tests. These tests are also called (3) _______ (4) _______ tests since the samples do not have to be drawn from normally distributed populations. Nonparametric inference tests have fewer requirements or assumptions

(1) parametric

(2) nonparametric

(3) distribution

(4) free

about (5) _______ characteristics. However,	(5) popula-tion
they are not used in place of parametric tests	
all the time because many parametric	
tests are (6) _______ with regard to violations	(6) robust
of underlying assumptions. The main advantages	
of (7) _______ tests are that they are more	(7) para-metric
(8) _______ and (9) _______ than	(8) powerful (9) versatile
nonparametric tests. As a general rule, an	
investigator will use (10) _______	(10) para-metric
tests whenever possible.	
The test most often employed with	
(11) _______ data is the nonparametric	(11) nominal
test called chi-square. It is symbolized	
(12) _______. With this type of data,	(12) χ^2
observations are grouped into several	
discrete, (13) _______ (14) _______	(13) mutually (14) exclusive
categories and one counts the (15) _______	(15) frequency

of occurrence in each category.	
The χ^2 test allows one to evaluate if the	
frequency observed in each category is	
significantly (16) _______ from the	(16) different
frequency (17) _______ in each category	(17) expected
if sampling were (18) _______	(18) random
from the (19) _______ (20) _______	(19) Null (20) Hypo- thesis
Population.	
In order to calculate χ^2 we must first	
determine the (21) _______ we would expect	(21) frequency
in each cell. This is symbolized	
by (22) _______. The frequency observed is	(22) f_e
symbolized (23) _______. The closer the	(23) f_o
(24) _______ frequency of each	(24) observed
(25) _______ is to the expected frequency for	(25) cell
that cell, the more reasonable is (26) _______.	(26) H_0
The greater the difference between	

(27) _______ and (28) _______, the

(27) f_o
(28) f_e

more reasonable

(29) _______ becomes.

(29) H_1

To calculate χ^2 the difference between

f_o and f_e is (30) _______ and divided by

(30) squared

(31) _______ and then

(31) f_e

(32) _______ over all of the (33) _______.

(32) summed
(33) cells

In equation form this is expressed:

$$\chi^2_{\text{obt}} = \sum \frac{((34)\_\_\_\_ - (35)\_\_\_\_)^2}{(36)\_\_\_\_}$$

(34) f_o
(35) f_e
(36) f_e

The summation is over (37) _______ the cells.

(37) all

In the case of a single variable experiment we calculate f_e based on information about the Null Hypothesis Population. For the experiment with a single variable there are (38) _______ degrees of freedom. The decision rule for evaluating the null hypothesis states:

(38) $k - 1$

If χ^2_{obt} (39) _______ χ^2_{crit}, reject H_0	(39) ≥
The χ^2 test is a (40) _______ test. Since	
each cell difference	(40) nondi-rectional
(41) _______ to the value of χ^2_{obt}, the critical region for rejection	(41) adds
always lies under the (42) _______ hand	(42) right
tail of the χ^2 distribution.	
One of the main uses of the χ^2 test is	
determining whether (43) _______ variables	(43) two
are related. This is done by using a	
(44) _______ table. This is a (45) _______ -	(44) contin-gency (45) two
way table showing the (46) _______	(46) contin-gency
between two variables where the variables	
have been classified into (47) _______	(47) mutually
(48) _______ categories and the cell	(48) exclusive
entries are (49) _______. The null hypo-	(49) frequencies
thesis states that there is no (50) _______	(50) contingency

between variables in the population. If H_1 is true then proportions on one of the variables should be (51) _______ for different categories of the other variable.

(51) different

The calculation of χ^2_{obt} for a contingency table uses the same formula as shown above. The most difficult task is to calculate the value of (52) _______ for each cell. Since we do not know the population proportions, we (53) _______ them from the (54) _______. The value of f_e can be found by (55) _______ the (56) _______ for that cell and dividing by (57) _______. The marginals are the (58) _______ and (59) _______ totals. If the calculation of f_e is correct then the row and column (60) _______ of f_e should equal the row and

(52) f_e
(53) estimate
(54) sample
(55) multiply-ing
(56) marginals
(57) N
(58) row
(59) column
(60) totals

column (61) _______. Once f_e is calculated then χ^2_{obt} can becalculated using the formula shown above.

(61) marginals

The evaluation of χ^2_{obt} is the same as in the one variable case except that there are (62) _______ degrees of freedom for χ^2_{crit}. This is because the degrees of freedom are equal to the number of (63) _______ scores that are free to (64) _______, while keeping the totals (65) _______. Again the decision rule is:

(62) $(r-1)(c-1)$

(63) f_o

(64) vary

(65) constant

If χ^2_{obt} (66) _______ χ^2_{crit}, reject H_0

(66) $\geq$

The basic assumption in using χ^2 is that there is (67) _______ between each observation recorded in the contingency table. It is also necessary that the value of

(67) independence

(68) _______ is at least 5 for the	(68) f_e
tables where *r* or *c* is greater than	
(69) _______. χ^2 can be used when data are	(69) 2
of (70) _______ type of scaling so long	(70) any
as the data are reduced to	
(71) _______ (72) _______ categories and	(71) mutually (72) exclusive
there are appropriate (73) _______.	(73) frequen- cies
The Wilcoxin signed ranks test is used	
with the (74) _______	(74) correlated
(75) _______ design with data that are of	(75) groups
at least (76) _______	(76) ordinal
scaling. It is more powerful than the	
(77) _______ test but less	(77) sign
powerful than the (78) _______ test for	(78) *t*
correlated groups. The statistic calculated	
is (79) _______. Determining this statistic	(79) T_{obt}
involves four steps:	

1. Calculate the (80) _______ between each (81) _______ of scores.

(80) difference
(81) pair

2. Rank the (82) _______ values of the difference scores from (83) _______ to (84) _______.

(82) absolute
(83) smallest
(84) largest

3. Assign the resulting ranks the (85) _______ of the difference score whose absolute value yielded that rank.

(85) sign

4. Compute the (86) _______ of the ranks separately for the positive and negative signed ranks. The (87) _______ sum is T_{obt}.

(86) sum
(87) lower

To evaluate T_{obt} the decision rule is:

If T_{obt} (88) _______ T_{crit}, reject H_0

(88) $\leq$

The Wilcoxin signed ranks test takes into account the (89) _______ (90) _______ of the difference

(89) rank
(90) order

scores not the actual (91) _______ of the	(91) magni-tude
difference scores.	
In calculating T_{obt} tied scores are possible.	
If the raw scores yield a difference of	
(92) _______ these scores are disregarded	(92) 0
and the overall (93) _______ is reduced	(93) N
by (94) _______. If ties occur in the difference	(94) 1
scores the ranks of these scores are	
given a value equal to the (95) _______	(95) mean
of the tied ranks.	
The assumptions underlying the Wilcoxin	
signed ranks test are that the scores within	
each pair must be of at least (96) _______	(96) ordinal
scaling and that the (97) _______	(97) difference
scores must also have at least (98) _______	(98) ordinal
scaling.	
The Mann-Whitney U test analyzes the	
(99) _______ between the two	(99) separa-

	tion
sets of (100) _______ scores and allows	(100) sample
us to determine the (101) _______ of getting	(101) probability
the obtained separation or even	
(102) _______ separation if both sets of	(102) greater
scores are (103) _______ samples from	(103) random
(104) _______ populations. The	(104) identical
(105) _______ the separation between the	(105) greater
two samples, the less likely it is that they are	
random samples drawn from the same	
(106) _______. The more (107) _______	(106) population (107) overlap
between the two sample distributions the more	
reasonably (108) _______ explains	(108) chance
the results.	
The value (109) _______ is the smaller	(109) U_{obt}
of two numbers that indicate the degree of	
separation between the two samples for	

the Mann-Whitney U test. (110) _______ is the larger of the two numbers indicating the same thing

(110) U'_{obt}

The following equations are used to compute U_{obt} *and* U'_{obt}:

$$\text{equation 1: } U_{obt} = n_1 n_2 + \frac{n_1(n_1+1)}{2} - R_1$$

$$\text{equation 2: } U_{obt} = n_1 n_2 + \frac{n_2(n_2+1)}{2} - R_2$$

where

n_1 = the number of scores in group = (111) _______

(111) 1

n_2 = the number of scores in group = (112) _______

(112) 2

R_1 = the (113) _______ of ranks in group 1

(113) sum

R_2 = the (114) _______ of ranks in group 2

(114) sum

Applying these equations to the following data, we obtain:

1 Control Group Score	Rank	2 Experimental Group Score	Rank	
28	(115)	32	(116)	(115) 1
30	______	34	______	(116) 3
35	(117)	42	(118)	(117) 2
	______	46	______	(118) 4
	(119)		(120)	(119) 5
	______		______	(120) 6
			(121)	(121) 7
			______	
R_1 = (122) ______		R_2 = (123) ______		(122) 8 (123) 20
n_1 = (124) ______		n_2 = (125) ______		(124) 3 (125) 4

Solving the equation for U_{obt}:

$$U_{obt} = n_1 n_2 + \frac{n_1(n_1 + 1)}{2} - R_1$$

$$U_{obt} = (126)\_\_(127)\_\_ + \frac{(128)\_\_ \times (129)\_\_}{2} - (1$$

(126) 3
(127) 4
(128) 3
(129) 4
(130) 8
(131) 10

Obtaining a value for U or U' is meaningless unless we know the (132) _______ of obtaining that value or any more (133) _______ assuming chance alone is at work. This information is available in Tables

(132) probability
(133) extreme

C_1-C_4. The tables have cells with

two entries. The upper entry is the

(134) _______ value of U for — (134) highest

various n_1 and n_2 combinations which will

allow (135) _______ — (135) rejection

of H_0. The lower entry is the lowest value
of (136) _______ — (136) U'

that will allow rejection of H_0 at a given

level of alpha.

Both U and U' measure the same

degree of (137) _______ — (137) separation

so one needs only to evaluate one or the

other. If we were evaluating the value of

U_{obt} using the above data at alpha =

$0.05_{1\text{ tail}}$, we would use table (138) _______. — (138) C_4

For $n_1 = 3$ and $n_2 = 4$, the largest value of

U that will allow rejection of H_0 is

(139) _______. Since the value of — (139) 0

U_{obt} is 2, we (140) _______ — (140) retain

H_0. If there had been tied scores in this problem we would simply assign the tied scores a value equal to the (141) _______ of the tied ranks.	(141) average
The (142) _______ test is used as a substitute for the independent groups, one-way	(142) Kruskal-Wallis
parametric ANOVA. It requires that the data be of (143) _______ scaling and does not require -	(143) ordinal
population (144) _______. It is a (145) _______	(144) normality (145) nonpara-metric
test. The statistic used is(146) _______.	(146) H_{obt}
Essentially, Kruskal-Wallis tests whether the sums of the (147) _______ for the *k* samples	(147) ranks
are so different that it is unreasonable to believe that the samples were randomly selected from populations with identical (148) _______.	(148) distribu-tions
To compute H_{obt}, all the scores are (149) _______	(149) combined
and rank (150) _______. Next the	(150) ordered

(151) _______ are summed for each sample. (151) ranks
It is these (152) _______ of ranks which are (152) sums

evaluated. The equation for computing H_{obt} is:

$$H_{obt} = \left[\frac{12}{N(N-1)}\right]\left[\frac{(153)\_\_}{(154)\_\_} + \frac{R_2^{\,2}}{n_2} + \frac{R_3^{\,2}}{n_3} + \cdots + \frac{R_k^{\,2}}{n_k}\right]$$

(153) $R_1^{\,2}$
(154) n_1

EXERCISES

In the problems below, if f_e has a decimal remainder, round it to one decimal place in calculating χ^2.

1. A designer of electronic equipment wants to develop a calculator which will have market appeal to high school students. Past marketing surveys have shown that the color of the numeric display is important in terms of market preference. The designer makes up 210 sample calculators and then has a random sample of students from the area high schools rate which calculator they prefer. The calculators are identical except for the color of the display. The results of the survey were that 96 students preferred red, 82 preferred blue, and 32 preferred green.

   a. State H_1 for this experiment.
   b. State H_0 for this experiment.
   c. What is the value of χ^2_{obt}
   d. What do you conclude using $\alpha = 0.01$?

2. One of the important assumptions underlying the use of parametric statistics is that the sample is randomly selected from a normally distributed population. Consider a sample of

$N = 500$. A sample mean and standard deviation is calculated and we find that the following is true. Between the mean and $-1s$ there are 150 scores. Between the mean and $+1s$ there are 130 scores. Between $-1s$ and $-2s$ there are 70 scores. Between $+1s$ and $+2s$ there are 82 scores. Beyond +2 standard deviations are 30 scores. Beyond −2 standard deviations are 38 scores.

a. If the population from which this sample was selected were normally distributed, what would the expected frequencies be for each cell in a sample of size 500?
b. Using $\alpha = .01$ what would you conclude about the population from which this sample was selected?

3. A family therapist in a hospital wanted to know if patients with a terminal illness wanted to be informed of their true medical condition. The therapist also wondered if a person age had an effect on their attitude. Because of ethical constraints the therapist asked a healthy sample of subjects who were visitors to the hospital whether they would wish to be told if they had a terminal illness. The age of the respondents was also recorded. The results are shown in the table below.

		Attitude		
		Wanted to be informed	Did not want to be informed	Not sure
Age	< 21	90	15	18
	22 - 35	56	24	19
	36 - 55	50	40	30
	> 55	47	60	50

a. Draw a table with the values of f_e in each of the appropriate cells.
b. What is the value of χ^2_{obt}
c. What is the value of χ^2_{crit} for $\alpha = 0.01$?
d. What do you conclude?

4. A neuropsychologist wants to determine if people who have a dominant right cerebral hemisphere differ from people with a dominant left cerebral hemisphere in their choice of either music or reading as a preferred activity. He surveyed 127 subjects with the following results.

	Dominant Hemisphere	
	Left	Right
Music	48	18
Reading	35	26

a. What is the value of χ^2_{obt}?
b. What is the value of χ^2_{crit} for $\alpha = 0.05$?
c. What do you conclude?
d. What type error might one be making?

5. A social scientist wants to know if education and socioeconomic status (SES) are independent. He collects the following data.

	Education			
	No high School	**High School**	**College**	**Graduate School**
High SES	12	19	30	13
Low SES	31	26	32	9

a What do you conclude using $\alpha = 0.05$?

6. Consider the following table.

		Variable X	
		Category 1	**Category 2**
Variable Y	**Category 1**	6	4
	Category 2	3	5

What is the appropriate statistical test to use to analyze this data if it were all nominal data.

7. A group of pain researchers want to test the hypothesis that different religious groups have different pain complaints. The following data were collected from a review of the patient charts from a hospital pain clinic.

	W	X	Y	Z
Low back pain	9	16	18	40
Headache pain	12	19	36	18
Gastrointestinal pain	20	18	16	21

   a. State H_1.
   b. State H_0.
   c. What do you conclude using $\alpha = 0.01$?

8. Given the following data:

		Eyes	
		Blue	**Brown**
Hair	**Blond**	36	29
	Brown	23	49

   a Do you think hair color and eye color are independent (use $\alpha = 0.01$)?
   b. What type of error might you be making?

9. A psychologist wants to investigate whether there might be a relationship between birth complications and the development of schizophrenia. In a longitudinal study she gathers the following data.

	Complicated birth	**Uncomplicated birth**
Schizophrenia	**43**	**15**
No Schizophrenia	**129**	**416**

   a. State H_1.
   b. State H_0.
   c. What do you conclude using $\alpha = 0.05$?

10. A new drug is supposed to be effective in reducing motion sickness in people who are prone to such illness. A group of subjects are given a placebo and taken for a ride in a car over a preplanned route. At the end of the trip the subjects are asked to rate their illness on a 20-point scale. A week later the same subjects are given the new drug and taken for an identical ride and asked to rate their degree of illness again. The data are:

Subject	Placebo	Drug
1	16	12
2	20	20
3	18	13
4	10	8
5	15	14
6	12	13
7	15	20
8	19	7
9	17	5
10	13	10

a. State H_1.
b. State H_0.
c. What is the value of T_{obt}?
d. What do you conclude using $\alpha = 0.05_{2\ tail}$?

11. A group of clients requesting marital therapy were given communication skills training and then rated by independent observers before and after therapy on their ability to resolve problems in a series of hypothetical conflict situations. The results are shown below. A higher score indicates better performance on the task.

Couple	Before	After
1	56	71
2	44	66
3	39	55
4	72	92
5	43	45
6	50	90
7	61	68
8	78	70

a. What is the value of T_{obt}?
b. What do you conclude using $\alpha = 0.05_{2\text{ tail}}$?

12. A political advisor believes that his candidate should not spend time addressing groups of voters who have a low opinion of him. The advisor reasons if they have a low opinion the voter won't change his mind anyway. To test this idea the advisor gets a group from an audience to rate the candidate before and after a speech. From this group he selects a sample of voters who initially rated the candidate poorly and then analyzes the effect of the speech. Here are the data. A higher rating indicates a higher opinion.

Subject	Before	After
1	11	12
2	8	1
3	4	13
4	5	6
5	15	4
6	3	1
7	6	7
8	9	6
9	12	2
10	14	0

a. What is the value of T_{obt}?
b. What do you conclude using $\alpha = 0.05_{1\ tail}$?
c. What type error might one be making?

13. An animal geneticist is trying to pick an appropriate species of fish for repopulating a lake. He wants to compare how long certain types of fish live. Species A is used for a control group and Species B serves as the experimental group. A random sample of fish from both species is drawn and the following ages are recorded for life span in months.

Species A	Species B
100	102
90	100
36	99
82	12
91	80
56	64
101	79
77	

a. State the nondirectional alternative hypothesis.
b. State the null hypothesis
c. Analyze the data with a nonparametric test. What do you conclude, using $\alpha = 0.05_{2\ \text{tail}}$?

14. Someone has told you that left-handed people have different spatial reasoning abilities than right-handed people. You are skeptical, so you decide to test the idea. You randomly select 15 people from your class and administer a spatial reasoning test to them. A higher score reflects better spatial reasoning. You obtain the following results.

Left-handed	Right-handed
70	81
85	80
60	50
92	95
82	93
65	85
	90
	75
	84

a. State the null hypothesis.
b. State the alternative hypothesis.
c. Assume the data do not allow analysis with the t test. Analyze the data with the most powerful alternative test. What is your conclusion using $\alpha = 0.05_{2\ \text{tail}}$?

15. The following data were collected in an independent groups experiment to test the effect of different levels of a drug on blood pressure (mmHg). Assume the data seriously violate the assumptions underlying parametric ANOVA. Therefore, you will have to use an alternative test to analyze the data.

Drug		
Level 1	Level 2	Level 3
83	79	90
82	86	80
82	72	70
85	73	65
87	70	78

a. What test will you use?
b. What is your conclusion? Use $\alpha = 0.01$

16. In an independent groups experiment, four wines are rated by individuals according to taste preference. The resulting data are shown below. The rating scale is from 1 to 20, with 20 representing the highest possible score. Assume the data preclude use of parametric ANOVA because of assumption violations. Analyze these results using an alternative test.

Wine 1	Wine 2	Wine 3	Wine 4
2	4	5	8
3	10	9	13
6	6	12	15
3	7	17	19
1	11	16	14
5	8	20	18

a. What test will you use?
b. Using $\alpha = 0.05$, what is your conclusion?

TRUE-FALSE QUESTIONS

T F 1. Nonparametric tests are generally more powerful than parametric tests.

T F 2. Anytime it is appropriate to use a nonparametric statistic it is appropriate to use a parametric statistic.

T F 3. A χ^2 test can only be applied to nominally scaled variables.

T F 4. In general, nonparametric tests have fewer requirements or assumptions about population characteristics than parametric tests do.

T F 5. As a general rule an investigator should use parametric tests whenever possible to help minimize the probability of making a Type II error.

T F 6. In a single variable χ^2 experiment there are $N - 1$ degrees of freedom.

T F 7. χ^2 is basically a measure of the overall discrepancy between f_e and f_o.

T F 8. In any specific case $f_o - f_e$ should equal zero if H_0 is true.

T F 9. To use the χ^2 test, the categories in the contingency table must always be mutually exclusive.

T F 10. The value of f_o can be found by multiplying the marginals for that cell, and dividing by N.

T F 11. If χ^2 is negative then H_0 must be false.

T F 12. In a 2 x 2 table there are $(r - 1)(c - 1)$ degrees of freedom.

T F 13. The Kruskal-Wallis test is a nonparametric alternate for parametric one-way ANOVA, independent groups design.

T F 14. In general, the Kruskal-Wallis test is as powerful as parametric ANOVA.

T F 15. The theoretical sampling distribution of χ^2 assumes that the distribution is discrete.

T F 16. In order to properly use the χ^2 test each cell should have a value of fe equal to or greater than 10.

T F 17. The sampling distribution of χ^2 is normally distributed.

T F 18. The Wilcoxin signed ranks test is used only with ordinal data.

T F 19. The sign test is less powerful than the Wilcoxin signed ranks test.

T F 20. Both the χ^2 test and the Wilcoxin signed ranks test are one-tailed tests.

T F 21. If the ranking has been done correctly for the Wilcoxin signed ranks test, the sum of the unsigned ranks should equal $n(n+1)/2$.

T F 22. If $T_{\text{obt}} \geq T_{\text{crit}}$, reject H_0.

T F 23. When using the Wilcoxin signed ranks test, if the raw scores are tied, the scores are disregarded and N is reduced by 1.

T F 24. The proper use of the Wilcoxin signed ranks test requires that both the raw scores and the difference between the raw scores be of at least ordinal scaling.

T F 25. The Mann-Whitney U test tests the difference between sample means.

T F 26. In an independent groups experiment involving two groups, if chance alone were operating one would expect a great deal of overlap between the two sets of scores.

T F 27. To use the Mann-Whitney U test, n_1 must equal n_2.

T F 28. Even though for a given experiment U_{obt} and U'_{obt} have different values, they still indicate the same degree of separation.

T F 29. The Kruskal-Wallis test is used with a correlated groups design.

T F 30. The Kruskal-Wallis test analyzes the difference between sample means.

T F 31. Both the Mann-Whitney *U* test and the Kruskal-Wallis test analyze differences between sums of ranks.

SELF-QUIZ

1. The χ^2 test can be used for variables with _______ scaling as long as the categories are mutually exclusive.

   a. nominal
   b. ordinal

c. interval
d. ratio
e. all the above

2. The _______ test is the most powerful test for a repeated measures design.

   a. sign
   b. *t*
   c. Wilcoxin signed ranks
   d. all the tests are equally powerful

3. Which of the following tests are parametric statistical tests:

   a. sign test
   b. chi-square test
   c. Wilcoxin signed ranks test
   d. none of the above

4. If an experiment using frequency data tested the preference for 6 brands of soup, there would be _______ degrees of freedom.

   a. 1
   b. $N - 1$
   c. 5
   d. 6

5. The value of χ^2_{obt} for the table below is _______. (Assume equal probabilities for f_e in each cell.)

12	20	16	18	30

a. 96.00
b. 9.42
c. 8.13
d. 13.28

6. The value of χ^2_{crit} for the data in question 5 with $\alpha = 0.01$ is _______.

a. 6.635
b. 15.086
c. 13.277
d. 11.668

7. The conclusion for the data in question 5 is _______.

a. reject H_0
b. reject H_1
c. retain H_0
d. retain H_1

8. The value of f_e for the cell in row *W*, column *A* is _______.

	A	B	C
W	16	25	64
X	28	32	30
Y	19	10	42

a. 24.9
b. 3.2
c. 16.0
d. 63.0

9. The value of χ^2_{obt} for the table in problem 8 is _______.

   a. 19.01
   b. 21.38
   c. 24.87
   d. 16.82

10. The value of χ^2_{crit} for the table in problem 8 with $\alpha = 0.05$ is _______.

    a. 13.277
    b. 7.779
    c. 3.841
    d. 9.488

11. The conclusion for the data in problem 8 is _______.

    a. reject H_0
    b. reject H_1
    c. retain H_0
    d. fail to accept H_1

12. For the following table there is(are) _______ degrees of freedom.

	X	Y
A	**60**	**30**
B	**18**	**40**

    a. $k - 1$
    b. 1

c. 2
d. 4

13. The value of χ^2_{obt} for the table in problem 12 is _______.

    a. 47.43
    b. 17.96
    c. 4.07
    d. 9.71

14. For a low value of df the χ^2 distribution is _______.

    a. normally distributed
    b. positively skewed
    c. negatively skewed
    d. none of the above

15. The value of T_{obt} for the following data is _______.

Subject	Pre	Post
1	100	98
2	62	80
3	80	75
4	75	74
5	90	80
6	85	71

    a. 21
    b. –6
    c. 15
    d. 6

16. If there are 16 subjects in a repeated measures design then the sum of the unsigned ranks equals _______.

a. 136
b. 68
c. 272
d. 32

17. If $T_{obt} = 12$ and $T_{crit} = 10$, one would ________.

a. reject H_0
b. retain H_0
c. accept H_0
d. reject H_1

18. The statistics used for the Mann-Whitney *U* test measure ________.

a. the mean differences between the two groups
b. the direction of the differences between pairs of scores
c. the power of the experiment
d. the separation between the two sets of scores

19. Consider the following set of scores: 81, 83, 84, 84, 87. What rank would you give to a score of 84?

a. 3
b. 3.5
c. 4
d. 4.5

20. For the following data, the value of U_{obt} (not U'_{obt}) is ________.

Control	Experimental
100	90
74	85
86	87
72	78
82	73
80	84
90	

a. 14.5
b. 20.5
c. 21.5
d. 27.5

21. The value of U_{crit} for the data in problem 20, using $\alpha = 0.05_{2\ tail}$ is _______.

a. 36
b. 5
c. 6
d. 30

22. The conclusion for the data in question 20 is _______.

a. reject H_0
b. reject H_1
c. retain H_0
d. retain H_1.

Questions 23 - 28 pertain to the following information.

In Chapter 15, we presented the data from an independent groups design and asked if it was appropriate to use parametric ANOVA. The data are presented again below. The correct answer was that it was not appropriate to use parametric ANOVA because of unequal n's and homogeneity of variance assumption violation.

Group 1	Group 2	Group 3
6	5	20
7	4	92
9	4	68
8	3	31
2	1	4
3		10
		82
		212

23 Is it possible to analyze the data with an alternate test?

a. yes
b. no

24 If your answer to question 23 is yes, what is the name of the test?

a. *t* test for independent groups
b. *F* test
c. Kruskal-Wallis
d. Mann-Whitney *U* test

25 H_{obt} = ________ .

a. 10.25
b. 10.63
c. 15.96
d. 5.96

26 What are the df?

a. 1
b. 2
c. 3
d. need more information

27 Using $\alpha = 0.05$, H_{crit} = ________.

a. 3.841
b. 7.815
c. 5.991
d. 7.824

28 What do you conclude?

a. retain H_0. There is no difference in the populations.
b. accept H_0. There is no difference in the populations.
c. reject H_0. At least one of the population means differs from at least one of the others.
d. reject H_0. At least one of the distributions differs from at least one of the others.

ANSWERS

Exercises: 1a. H_1: there is a difference in color preference for the calculators. **1b.** H_0: there is no difference in color preference for the calculators. **1c.** $\chi^2_{obt} = 32.34$ **1d.** $\chi^2_{crit} = 9.210$; therefore reject H_0.

2a. f_e is shown in parentheses.

< -2s	-2s and -1s	-1s and mean	+1s and mean	+1s and +2s	> +2s
38 (11.4)	70 (68.0)	150 (170.6)	130 (170.6)	82 (68.0)	30 (11.4)

2b. $\chi^2_{obt} = 107.50$; $\chi^2_{crit} = 15.086$; therefore reject H_0. It does not seem reasonable to assume that this sample was randomly selected from a normally distributed population.

3a. Table of f_e.

59.9	34.3	28.8
48.2	27.6	23.2
58.4	33.4	28.1
76.5	43.7	36.8

3b. $\chi^2_{obt} = 57.36$ **3c.** $\chi^2_{crit} = 16.812$ **3d.** Attitude and age group are not independent.

4a. $\chi^2_{obt} = 3.30$ **4b.** $\chi^2_{crit} = 3.841$ **4c.** $\chi^2_{obt} < \chi^2_{crit}$; therefore retain H_0. It appears as if preference for music or reading and hemisphere dominance are independent. **4d.** Type II

5a. $\chi^2_{obt} = 7.01$; $\chi^2_{crit} = 7.815$; therefore retain H_0. SES and education appear to be independent.

6. Fisher's exact probability test.

7a. H_1: religion and area of pain are not independent. **7b.** H_0: religion and area of pain are independent . **7c.** $\chi^2_{obt} = 25.14$; $\chi^2_{crit} = 16.812$; therefore reject H_0.

8a. $\chi^2_{obt} = 7.65$; $\chi^2_{crit} = 6.635$; therefore it appears as if hair and eye color are not independent. **8b.** Type I.

9a. H_1: birth complications and occurrence of schizophrenia are not independent. **9b.** H_0: birth complications and occurrence of schizophrenia are independent. **9c.** $\chi^2_{obt} = 65.49$; $\chi^2_{crit} = 3.841$; therefore reject H_0.

10a. H_1: the new drug affects self rating of motion sickness. **10b.** H_0: the new drug has no effect on self rating of motion sickness. **10c.** $T_{obt} = 8.0$ **10d.** $T_{crit} = 5$; therefore retain H_0.

11a. $T_{obt} = 3$ **11b.** $T_{crit} = 3$; therefore reject H_0.

12a. $T_{obt} = 13$ **12b.** $T_{crit} = 10$; therefore retain H_0. **12c.** Type II.

13a. One species of fish has a different life span from the other. **13b.** There is no difference in the life span of the two species of fish studied. **13c.** $U_{obt} = 27.5$, $U_{crit} = 10$; therefore retain H_0.

14a. There is no difference in performance on a spatial reasoning task between right and left handed people. **14b.** There is a difference in performance on a spatial reasoning task between right and left handed people. **14c.** $U_{obt} = 19.5$, $U_{crit} = 10$; therefore retain H_0.

15a. Kruskal-Wallis **15b.** $H_{obt} = 3.84$; $H_{crit} = 9.210$. Therefore, retain H_0.

16a. Kruskal-Wallis **16b.** $H_{obt} = 15.10$; $H_{crit} = 5.991$. Therefore, reject H_0. At least one of the distributions differs significantly from at least one of the others.

True-False: **1.** F **2.** F **3.** F **4.** T **5.** T **6.** F **7.** T **8.** F **9.** T **10.** F **11.** F **12.** T **13.** T **14.** F **15.** F **16.** T **17.** F **18.** F **19.** T **20.** F **21.** T **22.** F **23.** T **24.** T **25.** F **26.** T **27.** F 28. T **29.** F **30.** F **31.** T

Self-Quiz: **1.** e **2.** b **3.** d **4.** c **5.** b **6.** c **7.** c **8.** a **9.** b **10.** d **11.** a **12.** b **13.** b **14.** b **15.** d **16.** a **17.** b **18.** d **19.** b **20.** b **21.** c **22.** c **23.** a **24.** c **25.** b **26.** b **27.** c **28.** d.

Chapter 18

Review of Inferential Statistics

This chapter presents a concise review of hypothesis testing and inferential statistics. To outline or review the concepts covered in this chapter would be redundant. The exercises and questions to follow can serve as a self-review over this material. You might wish to use it as a practice examination. Be sure to follow good procedure by using the most powerful test that is appropriate for the information presented.

EXERCISES

1. Name two different pieces of information given by the sampling distribution of a statistic.

2. What are the three steps in empirically generating a sampling distribution?

3. In an ideal experiment, what should be the magnitudes of power and alpha?

4. Why is the *z* test for independent groups rarely used?

5. A neurologist believes that certain chemical toxins can affect the ability of nerves to conduct impulses. Eleven experimental animals have their nerve conduction velocity measured before and after exposure to the toxin. The following results are observed. The measures are in meters per second.

Animal	Before Exposure	After Exposure
1	45	40
2	42	36
3	39	40
4	50	48
5	46	42
6	44	38
7	38	38
8	41	49
9	40	42
10	47	42
11	46	41

a. State the nondirectional alternative hypothesis.
b. State H_0.
c. What do you conclude using $\alpha = 0.01_{2\ \text{tail}}$?
d. What type error might you be making?

6. Analyze the data from problem 5 using the most powerful nonparametric test appropriate for this design. What do you conclude?

7. A study was performed to see if cars actually get the gas mileage the EPA says. A sample of six cars was randomly selected from a manufacturer, tuned to specifications and then mileage tested using identical procedures as the EPA uses. The mean miles per gallon (mpg) from the sample was 27.8 mpg with a standard deviation of 1.9 mpg. The EPA states that the mileage should be 30.0 mpg. What would you

conclude using $\alpha = 0.05_{2 \text{ tail}}$? Assume a normally distributed population.

8. An insurance company wanted to know if a high school driver's education class reduced the number of accidents 16-year old drivers had. Eighteen high schools were selected. The number of accidents per 100 students was measured among students who had and had not taken driver's education. The results are shown below. Assume that the groups are independent.

With Drivers Education	Without Drivers Education
8	**9**
6	**11**
7	**8**
2	**9**
3	**8**
4	**16**
0	**10**
9	**7**
7	**9**

a. State the directional alternative hypothesis.
b. State the null hypothesis.
c. What do you conclude using $\alpha = 0.01_{1 \text{ tail}}$?
d. Estimate the size of the effect, using $\hat{\omega}^2$.
e. Construct the 95% confidence interval for ${}_1 - {}_2$. What conclusions can you draw from this interval?

9. Analyze the data from problem 8 using the most powerful nonparametric test appropriate. What do you conclude?

10. In an effort to determine if past marital history is related to present marital adjustment, a retrospective survey of public records reveals the following information for a 10-year period. Out of 400 males selected for the sample, 150 had been married 2 or more times at the outset of the 10-year period. Of these 150, 100 were divorced again during the 10 years. Of the 250 males who were in their first marriage at the outset of the 10-year period, 123 got divorced during the following 10 years. The data are summarized in the table below.

	Divorced in 10 year period	
	No	**Yes**
One Marriage	**127**	**123**
Two or More Marriages	**50**	**100**

Are past marital history and future marital success related? Use $\alpha = 0.01$. Round f_e to one decimal place.

11. A study was conducted to determine if a correlation exists between the unemployment rate in a community and the number of suicides over a period of 10 years. The correlation coefficient for the 10 pairs of scores was calculated with a resulting $r_{\text{obt}} = .656$.
    a. State the null hypothesis.
    b. Is this value of r_{obt} significant using $\alpha = 0.01_{\text{2 tail}}$?

12. A consumer testing laboratory wants to know if a sample of 20 boxes of breakfast cereal with a mean weight of 15.6 ounces could reasonably have been drawn from a population which is supposed to contain 16 ounces with a standard deviation of 0.6 ounces. What should the investigator conclude using $\alpha = 0.05_{\text{2 tail}}$? Assume the population is normally distributed.

13. A manufacturer of feed grain has three different products that it wishes to test. The question of interest is whether groups of test animals live longer eating certain products. The following data are the life span in months of groups of test chickens.

Product A	Product B	Product C
40	36	42
39	35	40
34	42	43
38	38	39
42	46	47
37	37	38
41	40	44
43	46	47

a. What is the alternative hypothesis?
b. What is the null hypothesis?
c. Using $\alpha = 0.05$, what do you conclude?

14. A high school senior wants to know if studying for the Scholastic Aptitude Test (SAT) produces higher grades on the test. She surveyed two groups of students who had taken the test the year before. She took data from 20 students all of whom had about the same high school GPA. The following data were gathered on the two groups of students.

No Study	Study
620	750
580	685
540	770
720	600
710	500
400	790
456	490
500	540
525	645
600	500

a. State the nondirectional null hypothesis.

b. What do you conclude using the *t* test and $\alpha = 0.05_{2\ \text{tail}}$?

c. If you felt the t test was inappropriate because you believed the SAT scores were only ordinal in scaling, what other test would be appropriate?

d. What would you conclude using that test and $\alpha = 0.05_{2\ \text{tail}}$?

15. A neuropsychologist knows that in the unanesthetized population the mean reaction time is 285 milliseconds. The population is normally distributed. A sample of 17 subjects that has undergone general anesthesia but no surgical procedure are given this same task 24 hours after receiving the anesthesia. The reaction time for this random sample of subjects was $\overline{X} = 327$ msec and $s = 22$ msec. Does the anesthesia affect performance on this task? Use $\alpha = 0.05_{2\ \text{tail}}$.

16. A cosmetic manufacturer wants to add a new perfume to its line of products. Five test fragrances are presented to a sample of women. The order is randomized. The preferences are shown in the following table.

Test Fragrance

A	B	C	D	E	
16	21	38	30	9	114

a. What is the nondirectional alternative hypothesis?
b. What do you conclude using $\alpha = 0.05$?

17. A nutritionist believes that large doses of vitamin C will affect the number and severity of the colds a person gets during a winter. Two groups of volunteers are given either a placebo or 1000 mg of vitamin C per day. At the end of the winter, each subject was asked to rate his/her health with regard to colds on a 20-point ordinal rating scale. The results are shown below. Higher scores represent better health. Assume the scores come from a highly skewed population.

Placebo	Vitamin C
16	12
5	6
11	18
14	14
19	4
10	15
16	17
3	20

a. State the nondirectional alternative hypothesis.
b. State H_0.
c. What do you conclude using $\alpha = 0.05_{2\text{ tail}}$?
d. What type of error might you be making?

18. A health educator believes that there is evidence to support the hypothesis that there is a positive relationship between the amount someone is over their ideal body weight and their serum cholesterol. She obtains the following data.

Excess Weight (lbs)	Serum Cholesterol (mg/100 ml)
1	125
5	100
8	130
10	200
15	280
30	320
60	380
20	250
12	260

a. What is the value of r_{obt}?

b. Is r_{obt} significantly more positive than 0 using $\alpha = 0.05_{1\ tail}$?

c. What type error might we be making?

19. A behavioral scientist wishes to compare the effectiveness of five different weight loss programs. Six subjects are randomly assigned to each of the programs and the weight loss per month is recorded.

Program				
1	2	3	4	5
6	10	7	14	0
8	11	6	16	5
7	12	8	8	6
5	11	10	15	2
3	8	8	15	2
9	5	13	18	3

a. State H_0.
b. What is the value of F_{obt}?
c. What do you conclude using $\alpha = 0.01$?
d. What is the size of the effect?
e. What does a planned comparison between groups 1 and 4 reveal? (Use $\alpha = 0.01_{\text{2 tail}}$)
f. Compare groups 2 and 5 using the most conservative post hoc test you know using $\alpha = 0.01$.

20. A pilot study was conducted in a school to see if a cancer information curriculum was effective in conveying new information to students. A pre- and post-test were administered to a small sample of students. The data are shown below.

Student	A	B	C	D	E	F	G	H	I	J
Pre-Test	56	58	73	65	60	62	64	68	70	73
Post-Test	75	81	72	80	69	68	69	64	69	89

a. What do you conclude when analyzing these data using the sign test with $\alpha = 0.05_{\text{1 tail}}$?
b. What do you conclude using $\alpha = 0.05_{\text{1 tail}}$ and the Wilcoxin signed ranks test?

c. What do you conclude using $\alpha = 0.05_{1\ \text{tail}}$ and the t test?
d. How do you explain these differing results?

21. What is the value of t_{obt} for evaluating the significance of $r =$ .83 for $N = 15$ pairs of scores?

22. A veterinarian believes his diet formula fed to dogs will produce healthier, less fat animals. His records collected on thousands of dogs show a mean of 22% fat with a standard deviation of 5.4. The data is normally distributed.
    a. If he tests his formula on a sample of 30 dogs, what is the power of this experiment to detect an effect of the diet such as to produce a mean decrease in fat of 2%? Use $\alpha = 0.05_{1\ \text{tail}}$.

    b. What number of dogs should he use to achieve a power of .8500 to detect the effect postulated in 22a?

23. A study was conducted to determine if people's performance on a standard task is related to their level of anxiety. A group of 1,018 subjects was randomly assigned to one of three conditions designed to create a low, moderate or high level of anxiety. Their performance on the standard task was rated as poor, fair, or good by an independent observer. The following results were obtained.

		Performance Rating		
		Poor	**Fair**	**Good**
Anxiety Level	**Low**	**130**	**100**	**80**
	Moderate	**47**	**193**	**162**
	High	**170**	**80**	**56**

a. State H_1.
b. State H_0.
c. What do you conclude using $\alpha = 0.01$? Round f_e to one decimal place accuracy.

24. The value of s_W^2 from a one-way ANOVA experiment that resulted in a significant overall F was 21.76. There were equal n's in each of four groups. The total degrees of freedom equaled 27. The means of the groups are as follows:

$$\overline{X}_1 = 17.2 \quad \overline{X}_2 = 14.8 \quad \overline{X}_3 = 21.3 \quad \overline{X}_4 = 24.0$$

a. Using the Newman-Keuls test and $\alpha = 0.05$, is group 1 different from group 3?
b. Using the Newman-Keuls test and $\alpha = 0.05$, is group 2 different from group 3?
c. Using the HSD test, is group 2 different from group 3?
d. Explain the difference between the conclusions of b and c.

25. Is it reasonable to conclude that the following random sample of IQ scores could have been drawn from a normally distributed population of scores with $\mu = 100$? Use $\alpha = 0.05_{2\ \text{tail}}$.

IQ Scores
107
106
92
103
110
105
100
116

26. A sleep researcher believes that when a person is sleep deprived for over 20 hours he becomes depressed. A group of 12 subjects is given a depression rating scale with ordinal scaling properties before and after 20 hours of sleep deprivation. The data are:

Subject	Before	After
1	56	65
2	70	71
3	40	43
4	37	59
5	65	70
6	50	70
7	55	61
8	72	65
9	68	64
10	66	80
11	61	69
12	57	71

Higher scores indicate increased depression.

a. State the nondirectional alternative hypothesis.
b. State the null hypothesis.
c. What do you conclude using $\alpha = 0.05$ Assume population depression scores are J-shaped

27. A study evaluates the effectiveness of three types of therapy in producing self-reliance. Fifteen volunteers are randomly assigned, five to each type of therapy. A questionnaire measuring self-reliance is administered after therapy is completed. The following results are obtained. Higher scores indicate greater self-reliance.

	Therapy	
Type A	**Type B**	**Type C**
10	9	16
11	13	18
12	14	19
15	8	17
7	14	15

a. Assume the data meet the assumptions of parametric ANOVA. What do you conclude, using $\alpha = 0.05$?
b. Using ^{2}, determine the size of the effect.
c. Assume the data seriously violate the assumptions of parametric ANOVA. Use another test to analyze the data. Again use $\alpha = 0.05$.

28. The researcher who did the study in question 27 has a hunch that the three types of therapy may affect males and females differently. Therefore a second study is undertaken to replicate the first study and in addition to see if there is a male/female difference in the effects of the therapy types. An independent groups design is conducted with 5 subjects per cell. The same questionnaire is used. The following data are collected.

Gender	Therapy Type A	Type B	Type C
Female	10	9	16
	11	13	18
	12	14	19
	15	8	17
	7	14	15
Male	9	8	18
	12	12	18
	12	15	17
	14	9	15
	8	14	20

What do you conclude, using $\alpha = 0.05$?

TRUE-FALSE QUESTIONS

T F 1. H_0 and H_1 must be mutually exclusive and exhaustive.

T F 2. If H_0 is true then H_1 must also be true.

T F 3. H_0 always states that the dependent variable has no effect on the independent variable.

T F 4. In a replicated measures design the Null Hypothesis Population has a population of difference scores with $\mu_D = 0$.

T F 5. The sampling distribution of a statistic can be generated by either an empirical or theoretical approach.

T F 6. The critical region for rejection is all the area under the curve.

T F 7. The critical value of a statistic depends on the alpha level.

T F 8. If one retains H_0 and H_0 is true, one has made a Type I error.

T F 9. $\alpha + \beta = 1$

T F 10. As power increases, alpha decreases.

T F 11. If in reality H_0 is true, then increasing power increases the probability of making a Type I error.

T F 12. As α increases, β decreases.

T F 13. With single sample experimental designs, one or more of the population parameters must be known.

T F 14. The *t* and *z* tests evaluate the effect of the independent variable on the sample mean while the ANOVA evaluates the effect of the independent variable on the sample variance.

T F 15. The sampling distribution of the mean has a mean of $\mu_{\overline{X}} = \mu$.

T F 16. The mean of all sampling distributions equals 0.

T F 17. If the raw score population is normally distributed, then so is the corresponding sampling distribution of the mean regardless of sample size.

T F 18. If the raw score population is not normally distributed, then the corresponding sampling distribution of the mean will not be normally distributed regardless of sample size.

T F 19. For a *t* test with single sample there are $N - 1$ degrees of freedom.

T F 20. The Wilcoxin signed rank test takes into account the magnitude and direction of the raw scores.

T F 21. To utilize the sign test, P must equal Q.

T F 22. Ties are not counted in the sign test.

T F 23. The Mann-Whitney U test requires equal n's in both groups.

T F 24. $F_{obt} = SS_B/SS_W$ (one-way ANOVA).

T F 25. If there are k treatments and k means between $\overline{X}_i$, and $\overline{X}_j$, then Q_{crit} will be the same for the HSD test and Newman-Keuls test when comparing $\overline{X}_i$ and $\overline{X}_j$.

T F 26. In the analysis of variance, it is not necessary for the sample variances to be homogeneous since this is what the statistic analyzes.

T F 27. The χ^2 test is only for use with nominal data.

T F 28. There are $(r - 1) + (c - 1)$ degrees of freedom in the χ^2 test.

T F 29. The χ^2 test can be used in a repeated measures design.

T F 30. For the proper use of the χ^2 test, f_e should equal f_o.

T F 31. To properly use the χ^2 test, the data must be reduced to mutually exclusive categories and appropriate frequencies.

T F 32. Power and sample size are directly related, while power and sample variance are inversely related.

T F 33. It is not possible to test the null hypothesis that $P = 0.50$.

T F 34. The Kruskal-Wallis test is a nonparametric test used in conjunction with a one-way independent groups design.

T F 35. In two-way ANOVA, we test for the main effects of two variables and their interaction.

T F 36. A significant interaction means the effects of one of the variables are not the same at all levels of the other variable.

SELF-QUIZ

1. The value of z_{crit} for $\alpha = 0.02_{2\text{ tail}}$ is _______.

   a. ±1.96
   b. ±1.64
   c. ±2.33
   d. ±2.58

2. The value of t_{crit} for $\alpha = 0.05_{2\text{ tail}}$ in a repeated measures experiment with $N = 20$ subjects is _______.

   a. ±2.093
   b. ±1.960
   c. ±2.861
   d. ±2.086

3. The value of r_{crit} for $\alpha = 0.01_2$ tail with $N = 15$ pairs of observations is _______.

   a. ±0.5923
   b. ±0.6055
   c. ±0.6226
   d. ±0.6411

4. The F distribution is _______.

   a. positively skewed
   b. has no negative values
   c. has a median approximately equal to 1
   d. all the above

5. There are _______ degrees of freedom for s_W^2
   a. $k - 1$
   b. $N - k$
   c. $N - 1$
   d. $N \times k$

6. The area under the critical region for rejection when $\alpha = 0.05_2$ tail is _______.

   a. 0.05
   b. 0.025
   c. 0.95
   d. 1.00

7. Which of the following levels of alpha is illegal?

   a. 0.01
   b. 0.05
   c. 0.10
   d. none of the above, all are permissible depending on the situation

8. If power equals 0.860, then β equals _______.

   a. 1.860
   b. 0.860
   c. 0.140
   d. 0.950

9. If $\mu = 37$, $\sigma = 56$, and $\overline{X} = 42.6$, then the value of z_{obt} using the *z* test for single samples equals _______.

   a. 1.00
   b. –1.00
   c. 0.68
   d. cannot be determined from information given

10. Consider an independent groups design with $k = 2$ and $s_1^2 = 0.75$ and $s_2^2 = 12.6$ and $n_1 = n_2 = 8$. What test should you use to analyze these data?

    a. *t* test
    b. *F* test
    c. Mann-Whitney *U* test
    d. any of the above

11. Which of the following do (does) not require equal *n*'s in each group?

    a. *t* test for independent groups
    b. ANOVA
    c. Mann-Whitney *U* test
    d. none of the above require equal *n*'s
    e. a and c

12. If $t_{obt} = 2.86$ for an independent groups design, then F_{obt} for the same data equals _______.

    a. 8.18
    b. 1.69
    c. 2.86
    d. cannot be determined from information given

13. Consider the following data from an independent groups design.

Condition 1	Condition 2
16	17
31	25
19	24
20	31
22	30
19	29
25	24

What is the value of t_{obt}?

    a. −2.34
    b. −1.53
    c. 2.34
    d. −1.69

14. If the data were ordinal what would the value of U_{obt} be from the data in problem 13?

    a. 35
    b. 17
    c. 14
    d. 49

Before	After
300	800
240	650
350	900
400	500
500	300
200	201
375	390
550	700
600	630

What is the value of t_{obt}?

a. −4.08
b. 1.46
c. −2.02
d. −3.52

16. If the data in problem 15 were analyzed using the Wilcoxin signed rank test, the value of T_{obt} would be ________.

a. 7
b. 29
c. 36
d. 6

17. If one were using the sign test to analyze the results of problem 15, the probability of getting the results observed or results more extreme would be ________.

a. 0.1406
b. 0.0392
c. 0.0703
d. 0.0899

18. What is the value of F_{obt} for the following data from an experiment with $k = 3$ groups and ratio scaled variables.

Group A	Group B	Group C
1.8	1.9	0.4
1.6	2.0	0.2
1.0	2.0	0.8
1.3	1.6	0.3

a. 2.20
b. 29.05
c. 6.49
d. 5.39

19. Referring to the data in problem 18, what is the value of Q_{crit} for comparing Group *A* and Group *C* using $\alpha = 0.05$ and the Newman-Keuls test?

a. 4.60
b. 3.95
c. 5.43
d. 3.20

20. Referring to the data in problem 18, what is the value of Q_{crit} for comparing Group *A* and Group *C* using $\alpha = 0.05$ and the HSD test?

a. 4.60
b. 3.95
c. 5.43
d. 3.20

21. What is the value of Q_{obt} for a post hoc comparison of groups A and *C* from problem 18?

    a. 7.26
    b. 0.14
    c. 1.00
    d. 10.88

22. A scientist asks a group of 75 other scientists which of three professional organizations they believe has done the most to advance the cause of free scientific investigation. The following data were obtained.

Organization			
A	**B**	**C**	
30	**25**	**20**	**75**

    What is the value of χ^2_{obt}

    a. 0.60
    b. 1.00
    c. 2.00
    d. 9.00

23. The following data are obtained on whether there is a relationship between type of physician and treatment preference for cancer. Round f_e to one decimal place accuracy.

	Treatment Preference		
	Surgery	Chemo-therapy	Both
Surgeon	62	27	46
Internist	37	89	47

What is the value of χ^2_{obt}?

a. 3.79
b. 16.03
c. 21.06
d. 35.21

24. What is the value of χ^2_{obt} for the following table which relates sex to birth control preference? Round f_e to one decimal place accuracy.

	Oral Contra-ceptive	Vasec-tomy
Male	50	10
Female	20	21

a. 12.04
b. 13.67
c. 8.71
d. 1.27

25. What is the value of t_{obt} for the test of significance of r_{obt} if $r_{obt} = .58$ and $N = 30$?

    a. 3.58
    b. 6.30
    c. 3.77
    d. 2.94

26. The following data are collected in a two-way independent groups design. Assume the data are ratio scaled and come from normal populations.

	Variable B	
Variable A	**Level 1**	**Level 2**
Level 1	2	7
	1	5
	5	8
	3	6
Level 2	9	1
	7	1
	8	2
	6	4

Which of the following conclusions is correct? Use $\alpha = 0.05$.

a. There are no significant effects
b. There is a significant main effect for variable *A*, and no other significant effects
c. There is a significant main effect for variable *B*, and no other significant effects
d. There is a significant interaction effect and no other significant effects

ANSWERS

Exercises: 1. First, all the values the statistic can take and second, the probability of getting each of those values if chance alone is responsible or sampling is random from the Null Hypothesis Population.

2. Step 1: all possible different samples of size N that can be formed from the population are determined. Step 2: the statistic for each of the samples is calculated. Step 3: the probability of getting each value of the statistic is calculated under the assumption that sampling is random from the Null Hypothesis Population.

3. Ideally, power would be high (power = 1.0000) and alpha would be low ($\alpha = 0$).

4. Because the experimenter rarely knows the population parameters μ and σ.

5a. H_1: Exposure to the toxin affects nerve conduction velocity. **5b.** H_0: Exposure to the toxin has no effect on nerve conduction velocity. **5c.** $t_{obt} = 1.51$; $t_{crit} = \pm 3.169$. Therefore, retain H_0. **5d.** Type II.

6. Using the Wilcoxin signed ranks test, $T_{obt} = 13.5$; $T_{crit} = 3$; therefore retain H_0.

7. $t_{obt} = -2.84$; $t_{crit} = \pm 2.571$. Therefore reject H_0. It is not reasonable to assume that this sample was drawn from a population with $\mu = 30$.

8a. H_1: driver's education classes reduce the number of accidents in which students are involved. **8b.** H_0: driver's education classes do not reduce the number of accidents in which students are involved. **8c.** $t_{obt} = -3.40$; $t_{crit} = -2.583$. Therefore, reject H_0. **8d.** $\hat{\omega}^2 = 0.370$. Driver's education classes account for 37.0% of the variance in the number of accidents. **8e**. The 95% confidence interval = -7.39 – -1.72.

9. Using the Mann-Whitney *U* test, $U_{obt} = 7.5$ and $U'_{obt} = 73.5$; $U_{crit} = 14$. Therefore, reject H_0.

10. $\chi^2_{obt} = 11.59$; $\chi^2_{crit} = 6.635$. Therefore, reject H_0. Past marital history and current adjustment are not independent.

11a. H_0: the sample is a random sample from a population with $\rho = 0$ and that any correlation in the sample is due to chance alone. **11b.** $r_{obt} = 0.656$; $r_{crit} = \pm 0.7646$. Therefore, retain H_0.

12. $z_{obt} = -2.98$; $z_{crit} = \pm 1.96$. Therefore, it is not likely that the sample boxes were drawn from a population with the stated parameters (i.e., reject H_0).

13a. H_1: at least one of the products differentially affects life span. **13b.** H_0: none of the products differentially affects life span. **13c.** $F_{obt} = 1.79$; $F_{crit} = 3.47$. Therefore, retain H_0.

14a. H_0: studying has no effect on SAT scores. **14b.** $t_{obt} = -1.25$; $t_{crit} = \pm 2.101$. Therefore, retain H_0. **14c.** Mann-Whitney *U* test **14d.** $U_{obt} = 36$; $U'_{obt} = 64$; $U_{crit} = 23$. Therefore, retain H_0. Notice again that it is unlikely that a nonparametric test will allow us to reject H_0 if a parametric on the same data has not allowed rejection of H_0.

15. $t_{obt} = 7.87$; $t_{crit} = \pm 2.120$. Therefore, reject H_0.

16a. H_0: there is no difference in preference among the test fragrances. **16b.** $\chi^2_{obt} = 22.93$; $\chi^2_{crit} = 9.488$. Therefore, reject H_0.

17a. H_1: vitamin C affects the health ratings of individuals who take it. **17b.** H_0: vitamin C has no effect on the health ratings of individuals who take it . **17c.** We have used the Mann-Whitney *U* test because the normal assumption for the *t* test is violated and *N* is relatively small. In addition, some researchers would also argue that you can't use the *t* test because it requires interval or ratio data. $U_{obt} = 25.5$; $U_{crit} = 13$. Therefore, retain H_0. **17d.** Type II.

18a. $r_{obt} = 0.86$ **18b.** $r_{crit} = 0.5822$. Therefore, reject H_0. It seems likely that this sample was not drawn from a population of scores with $\rho = 0$. **18c.** Type I

19a. H_0: there is no difference in weight loss between these groups **19b.** $F_{obt} = 15.47$ **19c.** $F_{crit} = 4.18$. Therefore, reject H_0. **19d.** $\hat{\omega}^2 = 0.659$. The weight loss programs account for 65.9% of the variance. **19e.** $t_{obt} = 5.32$; $t_{crit} = \pm 2.787$. Reject H_0. These groups are not very likely to have been drawn from the same population. **19f.** Use HSD test. $Q_{obt} = 6.11$; $Q_{crit} = 5.15$. Therefore, reject H_0.

20a. p = .1719 and $\alpha = 0.05$. Therefore, retain H_0. **20b.** $T_{obt} = 6$, $T_{crit} = 10$. Therefore, reject H_0. **20c.** $t_{obt} = 2.97$; $t_{crit} = 1.833$. Therefore, reject H_0. **20d** The sign test was not powerful enough to allow us to reject H_0

21. $t_{obt} = 5.37$.

22a. Power = .6480 **22b.** $N = 53$.

23a. H_1: anxiety level and performance are not independent. **23b.** H_0: anxiety level and performance are independent. **23c.** $\chi^2_{obt} = 161.63$; $\chi^2_{crit} = 13.277$. Therefore, reject H_0.

24a. $Q_{obt} = 2.33$; $Q_{crit} = 2.92$. Therefore, we cannot say groups 1 and 3 are drawn from different populations. **24b.** $Q_{obt} = 3.69$; $Q_{crit} = 3.53$. Therefore, using this test, we reject H_0. **24c.** $Q_{obt} = 3.69$; $Q_{crit} = 3.90$. Therefore, retain H_0. **24d.** HSD is more conservative regarding Type I error probability than Newman-Keuls.

25. $t_{obt} = 1.95$; $t_{crit} = \pm 2.365$. Therefore, retain H_0. It is reasonable that this sample could have been drawn from a population with $\mu = 100$.

26a. H_1: sleep deprivation affects depression rating scores. **26b.** H_0: sleep deprivation has no effect on depression rating scores. **26c.** We have used the Wilcoxin signed ranks test instead of the t test because the population scores are not normally distributed and N is small. In addition, many researchers would rule out use of the t test because the data are only of ordinal scaling. $T_{obt} = 9$; $T_{crit} = 13$. Therefore, reject H_0.

27a. $F_{obt} = 8.49$; $F_{crit} = 3.74$. Therefore reject H_0. **27b.** $^{2} = 0.586$ **27c.** $H_{obt} = 9.04$; $H_{crit} = 5.991$. Therefore reject H_0.

28. F_{obt} (row) = 0.05; $F_{crit} = 4.26$. Retain H_0. F_{obt} (column) = 19.14; $F_{crit} = 3.40$. Reject H_0. F_{obt} (row × column) = 0.05; $F_{crit} = 3.40$. Retain H_0. There is a significant main effect for the therapy types, and no other significant effects. This study replicates the previous finding and extends it to males.

True-False: **1.** T **2.** F **3.** F **4.** T **5.** T **6.** F **7.** T **8.** F **9.** F **10.** F **11.** F **12.** T **13.** T **14.** F **15.** T

16. F **17.** T **18.** F **19.** T **20.** T **21.** F **22.** T **23.** F **24.** F **25.** T **26.** F **27.** F **28.** F **29.** F **30.** F **31.** T **32.** T **33.** F **34.** T **35.** T **36.** T.

Self-Quiz: **1.** c **2.** a **3.** d **4.** d **5.** b **6.** a **7.** d **8.** c **9.** d **10.** c **11.** d **12.** a **13.** b **14.** c **15.** c **16.** d **17.** b **18.** b **19.** d **20.** b **21.** a **22.** c **23.** d **24.** b **25.** c. **26.** d.